DISTRIBUTED SITUATION AWARENESS

For Kerri, Holli and Lachlan

Distributed Situation Awareness
Theory, Measurement and Application to Teamwork

PAUL M. SALMON
Monash University, Australia

NEVILLE A. STANTON
University of Southampton, UK

GUY H. WALKER
Heriot-Watt University, UK

&

DANIEL P. JENKINS
Sociotechnic Solutions Ltd, UK

CRC Press
Taylor & Francis Group
Boca Raton London New York

CRC Press is an imprint of the
Taylor & Francis Group, an **informa** business

CRC Press
Taylor & Francis Group
6000 Broken Sound Parkway NW, Suite 300
Boca Raton, FL 33487-2742

First issued in paperback 2017

© 2009 by Paul M. Salmon, Neville A. Stanton, Guy H. Walker and Daniel P. Jenkins
CRC Press is an imprint of Taylor & Francis Group, an Informa business

No claim to original U.S. Government works

Version Date: 20160226

ISBN 13: 978-1-138-07385-2 (pbk)
ISBN 13: 978-0-7546-7058-2 (hbk)

**Visit the Taylor & Francis Web site at
http://www.taylorandfrancis.com**

**and the CRC Press Web site at
http://www.crcpress.com**

Contents

List of Figures

List of Tables

Acknowledgements

The Human Factors Integration Defence Technology Centre is a consortium of defence companies and Universities working in cooperation on a series of defence related projects. The consortium is led by Aerosystems International and comprises Birmingham University, Brunel University, Cranfield University, Lockheed Martin, MBDA, and SEA. The consortium was recently awarded The Ergonomics Society President's Medal for work that has made a significant contribution to original research, the development of methodology, and application of knowledge within the field of ergonomics.

Aerosystems International	Birmingham University	Brunel University	Cranfield University
Dr Karen Lane	Prof Chris Baber	Professor Neville Stanton	Dr Don Harris
Dr David Morris	Professor Bob Stone	Dr Guy Walker	Andy Farmilo
Linda Wells	Dr Huw Gibson	Dr Daniel Jenkins	Geoff Hone
Kevin Bessell	Dr Robert Houghton	Dr Paul Salmon	Jacob Mulenga
Kelly Maddock-Davies	Richard McMaster	Amardeep Aujla	Ian Whitworth
Nicola Gibb	Dr James Cross	Kirsten Revell	John Huddlestone
Robin Morrison	Robert Guest	Laura Rafferty	Antoinette Caird-Daley

Lockheed Martin UK	MBDA Missile Systems	Systems Engineering and Assessment (SEA) Ltd
Mick Fuchs	Dr Carol Mason	Dr Anne Bruseberg
Lucy Mitchell	Grant Hudson	Dr Iya Solodilova-Whiteley
Mark Linsell	Chris Vance	Mel Lowe
Ben Leonard	Steve Harmer	Ben Dawson
Rebecca Stewart	David Leahy	Jonathan Smalley
		Georgina Fletcher

We are grateful to DSTL who have managed the work of the consortium, in particular to Geoff Barrett, Bruce Callander, Jen Clemitson, Colin Corbridge, Roland Edwards, Alan Ellis, Jim Squire, Alison Rogers and Debbie Webb.

This work from the Human Factors Integration Defence Technology Centre was part-funded by the Human Sciences Domain of the UK Ministry of Defence Scientific Research Programme. Further information on the work and people that comprise the HFI DTC can be found on www.hfidtc.com.

We would like to thank to the various participants who were involved in the case studies described. Special thanks go to all of the staff at the energy distribution company involved in the study described in Chapter 6, the tri-service staff involved in the MultiNational experiment described in Chapter 7, the staff at C2DC involved in the study described in Chapter 8 and the Brigade and BattleGroup staff involved in the study described in Chapter 9.

Special thanks go also to the various analysts who have been involved at some point in collecting the data used during this research. Whilst the end analysis was undertaken by the authors, a huge effort was involved in collecting the data used. Many thanks go to Laura Rafferty, Darshna Ladva, Kirsten Revell, Dr Chris Baber, Dr Robert Houghton and Richard McMaster for all their help in this respect.

Many thanks go also to a number of subject matter experts who have gone over and above the call of duty to review and refine analyses, discuss ideas and provide invaluable domain knowledge, opinions and viewpoints. Special mention goes to Major Mike Forster for his input into the digitised mission support system analyses described in Chapters 8 and 9.

About the Authors

Dr Paul M. Salmon
Human Factors Group, Monash University Accident Research Centre, Building 70,
Clayton Campus, Monash University, Victoria 3800, Australia
paul.salmon@muarc.monash.edu.au

Paul Salmon is a Senior Research Fellow within the Human Factors Group at the Monash University Accident Research Centre and holds a BSc in Sports Science, an MSc in Applied Ergonomics and a PhD in Human Factors. Paul has over seven years experience in applied human factors research in a number of domains, including the military, aviation, and rail and road transport and has worked on a variety of research projects in these areas. This has led to Paul gaining expertise in a broad range of areas, including situation awareness, human error and the application of human factors methods, including human error identification, situation awareness measurement, teamwork assessment, task analysis and cognitive task analysis methods. His current research interests include the areas of situation awareness, human error and the application of human factors methods in sport. He has authored and co-authored various scientific journal articles, conference articles, book chapters and books and was recently awarded the 2006 Royal Aeronautical Society Hodgson Prize for a co-authored paper in the society's *Aeronautical Journal*. Paul and his colleagues from the HFI DTC consortium were also awarded the Ergonomics Society's President's Medal in 2008.

Professor Neville A. Stanton
Transportation Research Group, University of Southampton
School of Civil Engineering and the Environment
Highfield, Southampton, SO17 1BJ
neville.stanton@soton.ac.uk

Professor Neville A. Stanton holds a Chair in Human Factors in the School of Civil Engineering and the Environment at the University of Southampton. He has published over 140 peer-reviewed journal papers and 14 books on Human Factors and Ergonomics. In 1998 he was awarded the Institution of Electrical Engineers Divisional Premium Award for a co-authored paper on Engineering Psychology and System Safety. The Ergonomics Society awarded him the Otto Edholm medal in 2001 and The President's Medal in 2008 for his contribution to basic and applied ergonomics research. In 2007 The Royal Aeronautical Society awarded him the Hodgson Medal and Bronze Award with colleagues for their work on flight deck safety. Professor Stanton is an editor of the journal Ergonomics and on the editorial boards of *Theoretical Issues in Ergonomics Science* and the *International Journal of Human Computer Interaction*. He is a Fellow and Chartered Occupational Psychologist registered with the British Psychological Society, and a Fellow of the Ergonomics Society. Professor Stanton has a BSc (Hons) in

Occupational Psychology from the University of Hull, an MPhil in Applied Psychology and a PhD in Human Factors from Aston University in Birmingham.

Dr Guy H. Walker
School of the Built Environment, Heriot-Watt University, Edinburgh, EH14 4AS, UK.

Guy Walker read for a BSc (Hons) degree in Psychology at Southampton University, specialising in engineering psychology, statistics and psychophysics. During his undergraduate studies he also undertook work in auditory perception laboratories at Essex University and the Applied Psychology Unit at Cambridge University. After graduating in 1999 he moved to Brunel University, gaining a PhD in Human Factors in 2002. His research focused on driver performance, situational awareness and the role of feedback in vehicles. Since this time he has worked for a human factors consultancy on a project funded by the Rail Safety and Standards Board, examining driver behaviour in relation to warning systems and alarms fitted in train cabs. Currently he works within the DTC HFI at Brunel University, engaged primarily in work on future C4i systems. He is also author of numerous journal articles and book contributions.

Dr Daniel P. Jenkins
Sociotechnic Solutions Ltd, St Albans, Herts, AL1 2LW, UK.

Dan Jenkins graduated in 2004, from Brunel University, with an MEng (Hons) in Mechanical Engineering and Design, receiving the 'University Prize' for the highest academic achievement in the school. As a sponsored student, he finished university with over two years experience as a design engineer in the automotive industry. Upon graduation, he went to work in Japan for a major car manufacturer, facilitating the necessary design changes to launch a new model in Europe. In 2005, Dan returned to Brunel University taking up the full-time role of Research Fellow in the Ergonomics Research Group, working primarily on the Human Factors Integration Defence Technology Centre (HFI-DTC) project. Dan studied part-time on his PhD in human factors and interaction design, graduating in 2008, receiving the 'Hamilton Prize' for the Best Viva in the School of Engineering and Design. Both academically and within industry he has always had a strong focus on human factors, system optimisation and design for inclusion. He has authored and co-authored numerous journal paper, conference articles, book chapters and books. Dan and his colleagues on the HFI DTC project were awarded the Ergonomics Society's President's Medal in 2008.

Commonly Used Analysis Acronyms and Initialisms

The following is a reference list of analysis acronyms and initialisms used within the document.

ATC	Air traffic control
AA	Avenues of approach
AP	Authorised person
BAE	Battlefield area evaluation
CAST	Coordinated awareness of situations by teams
CARS	Crew awareness rating scale
CCIRs	Commander's critical information requirements
CCO	Central command operator
CDM	Critical decision method
CE	Combat estimate
CG	Command group
COCR	Central operations control room operator
CONOPS	Concept of operations
CP	Competent person
C-SAS	Cranfield situation awareness scale
CSE	Cognitive systems engineering
CSSO	Combat service support operations
CTF	Coalition Task Force
DIME	Diplomatic, information, military and economic
DPs	Decision points
DSA	Distributed situation awareness
DSO	Decision support overlay
DSOM	Decision support overlay matrix
EAST	Event analysis of systemic teamwork
EBA	Effects based assessment
EBE	Effects based execution
EBO	Effects based operations
EBP	Effects based planning
GPRs	Global prototypical routines
HEI	Human error identification
HF	Human factors
HFI-DTC	Human Factors Integration Defence Technology Centre
HESH	High Explosive Squash Head
HQ	Headquarters
HRA	Human reliability analysis

HTA   Hierarchical task analysis
IDP   Internationally displaced persons
ISTAR   Information surveillance target acquisition and reconnaissance
IWS   Information workspace
KBD   Knowledge based development
KM   Knowledge management
LSSRs   Local state specific routines
KS   Knowledge support
MARS   Mission awareness rating scale
MAs   Manoeuvre areas
MCs   Mobility corridors
MNE4   MultiNational Experiment 4
MNIG   MultiNational Interagency Group
MoD   Ministry of Defence
MOUT   Military operations in urban terrain
MPS   Mission planning system
NAIs   Named areas of interest
NCW   Network centric warfare
NEC   Network enabled capability
NASA TLX   NASA Task Load Index
NGOs   National government organisations
OLP   Overhead line party
PSAQ   Participant SA questionnaire
QUASA   Quantitative Assessment of Situation Awareness
RFIs   Requests for information
SA   Situation awareness
SABARS   Situation Awareness Behavioural Anchored Rating Scales
SACRI   Situation Awareness Control Room Inventory
SAGAT   Situation awareness global assessment tool
SAP   Senior authorised person
SARS   Situation Awareness Rating Scales
SART   Situation Awareness Rating Technique
SHERPA   Systemic Human Error Reduction and Prediction Approach
SGT   Super grid transformer
SME   Subject matter expert
SPAM   Situation-present assessment method
SOI   Standard operating instructions
SA-SWORD   Subjective workload dominance metric
TAIs   Target areas of interest
UN   United Nations
VPA   Verbal protocol analysis

Chapter 1

Introduction

Introduction

A Challenger II tank, engaged in defending a bridge over the Shatt al Basra canal on the western outskirts of Basra, fires upon what its commander believes are enemy personnel moving in and out of an ammunitions bunker. Unbeknown to the commander the engaged target is actually two friendly Challenger II tanks from another squadron, sited in an over watch position adjacent to a dam only 1,500 metres to the southeast of his own position. The first High Explosive Squash Head (HESH) round fired lands short but the effects of the blast are sufficient to throw the crew members from the tanks' turrets. The second round is a direct hit, detonating in the commander's hatch of one of the Challenger tanks, killing its two occupants instantly.

The incident described is an extreme example of what can happen in a complex sociotechnical system when the operators working within the system are not fully cognisant of everything that they need to know. In the aftermath, the official government inquiry (Ministry of Defence, 2004) identified various causal factors, including a lack of what it called Situation Awareness (SA) on behalf of those involved. SA is the term that is used within Human Factors (HF) and ergonomics circles to describe the level of awareness that people have of the situation that they are engaged in; it focuses on how people develop and maintain a sufficient understanding of 'what is going on' (Endsley, 1995a) and what is likely to go on in order to achieve success in task performance.

Safe and efficient task performance within complex sociotechnical systems depends on operators acquiring and maintaining appropriate levels of SA. Systems, devices and procedures therefore need to be designed so that they facilitate, rather than inhibit, SA acquisition and maintenance. Designing systems in this manner depends on the accurate description of how SA operates in the system in question, of exactly what information SA comprises during task performance and of how this information is integrated and used by different human operators working within the system. Further, reliable and valid approaches for modelling SA are required in order to determine how new system, device, training programme and procedural designs affect SA during operations.

The concept is therefore a critical consideration in collaborative system design (Endsley et al., 2003; Salmon et al., 2008; Shu and Furuta, 2005). However, despite over two decades of research in the area, SA remains ambiguous; there is still huge debate over what SA actually is, what it comprises, what factors have an impact on it, how it is best acquired, how it can be measured and how its acquisition and maintenance can be supported through system design. The inescapable conclusion arrived at by the authors as a result of comprehensive reviews of SA theory (Salmon et al., 2008) and SA measures (Salmon et al., 2006) was that there remain significant issues that require resolution. Of most relevance, given the complexity of modern day sociotechnical systems and the increasing presence of teams (Fiore et al., 2003), are the limitations of existing models and measures when applied to teamwork, and many researchers working in the area have articulated the need for a greater understanding of SA in collaborative environments (e.g. Artman and Garbis, 1998; Gorman et al., 2006; Patrick et al., 2006; Salmon et al., 2006, 2008; Stanton et al., 2006, 2009; Shu and Furuta, 2005; Siemieniuch and Sinclair, 2006; Sonnenwald et al., 2004).

This book aims to address how SA operates within complex collaborative environments. For this purpose we present the results of our investigations, undertaken as part of the HFI-DTC research project, into the concept of SA within complex command and control environments. Specifically, the aims of our research were to investigate how current theoretical perspectives on SA relate to the concept of team SA, to explore and extend a new systems approach to describing SA in collaborative environments and to extend and validate a new approach to modelling and assessing SA in collaborative environments. The ultimate goal of this research from the outset was to formulate a model of SA in complex collaborative systems, develop and validate a new approach for modelling and evaluating SA during real world collaborative tasks, and to begin to postulate a set of guidelines for enhancing SA acquisition and maintenance through system design.

Research Activities Undertaken

The following activities were undertaken as part of our research:

1. *Literature reviews.* Comprehensive reviews of the academic literature surrounding SA and SA measurement were undertaken in order to evaluate and compare and contrast existing SA models and measurement approaches.
2. *Case studies.* Four case studies were undertaken as part of this research. The studies involved observing and analysing real world collaborative tasks undertaken in the civil and military domains in order to investigate the concept of team SA and its measurement.
3. *Synthesis of findings.* The findings from the literature reviews, experimental study and naturalistic studies were used to develop a model of SA for complex collaborative systems and to propose a series of initial guidelines for enhancing SA in collaborative systems.

Structure of the Book

This book has been constructed so that readers new to this specific subject area can read it linearly. An attempt has also been made to construct the individual chapters and case studies so that they can be read non-linearly, or independently of the early chapters, by those well versed in the area with an interest in the extensions and proposed developments documented in this book. This means that there is some repetition in the description of Distributed Situation Awareness (DSA) and its sub-concepts and also some of the findings from previous chapters, in an attempt to create a resource for experts. A brief summary of each chapter is presented below.

Chapter 2: What Really is Going on? Situation Awareness Literature Review

In order to set the scene and orient the reader, the second chapter presents an introduction to SA and then the findings derived from a comprehensive review of the literature on what is currently known about SA in complex systems. The most prominent models of individual and team SA are described and then compared and contrasted. In conclusion, the flaws associated with current SA theory are discussed and the requirement for further investigation in the area, particularly with regard to SA in complex collaborative systems, is outlined.

Chapter 3: How Do We Know What They Know? Situation Awareness Measurement Methods Review

Analysing SA in any domain requires the provision of valid and reliable SA measurement approaches. The third chapter introduces the concept of SA measurement and presents a comprehensive review and evaluation of existing SA measurement approaches. Over 20 SA measurement approaches are compared and contrasted using a set of HF methods criteria and the advantages and disadvantages of each approach are discussed. In conclusion, the lack of suitable measures for assessing SA during collaborative real world tasks is discussed and the requirement for a new team SA measure is articulated.

Chapter 4: Distributed Situation Awareness: A New View on SA in Collaborative Systems and its Measurement

The inadequacies of existing SA theory and measures discussed in Chapters 2 and 3 highlight the need for new approaches in the area of team SA. In the fourth chapter, a recently developed model of SA in complex collaborative systems – the DSA model – is presented, along with a new approach for modelling and assessing SA in such environments, the propositional network methodology. Both approaches are outlined using examples and the relative merits of each approach are discussed.

Chapter 5: Distributed Situation Awareness in the Real World: A Case Study in the Energy Distribution Domain

The next four chapters focus on case studies of DSA in complex collaborative environments. The first of the case studies, focusing on DSA in the energy distribution domain, is presented in Chapter 5. The energy distribution case study involved the use of the propositional network methodology to analyse DSA during three energy distribution maintenance scenarios. The findings from the case study are discussed in relation to DSA theory, including the structure, quality and content of DSA in each case. In particular, the concept of compatible SA is focused on and its implications for DSA theory are discussed.

Chapter 6: Distributed Situation Awareness in Military Network Enabled Capability Systems: MultiNational Experiment 4

At this point the book moves into the military domain and focuses on the impact that newly developed Network Enabled Capability (NEC) based systems are likely to have on DSA during operational activities. Chapter 6 presents a case study on DSA during the MultiNational Experiment 4 (MNE4), which was undertaken in order to test a new approach (effects based operations) and a new technological system (Information workspace), both of which were designed to support modern day multinational warfare operations. The findings are discussed in relation to their implications for DSA theory and for the design of future electronic collaborative warfare systems.

Chapter 7: Out with the Old and In with the New: A Comparison of Distributed Situation Awareness using Analogue and Digital Mission Support Systems

Staying within the military domain but moving into the realm of land warfare, Chapter 7 presents a comparison of DSA during land warfare mission planning activities when using a traditional paper map planning process and a newly developed digital mission support system. Digitised systems are currently being developed to support future warfare activities and it is claimed that they will lead to enhanced levels of SA in teams (a claim that does not yet appear to have been corroborated by valid scientific means). Training scenarios involving planning activities using both approaches (old paper map and new digitised system) were analysed using the propositional network approach. The findings derived from this study are discussed in terms of the impact that the new electronic mission planning system had on DSA during the planning activities observed.

Chapter 8: Is it Really Better to Share? Analysis of a New Electronic Mission Support System and Implications for System Design

Chapter 8 presents a second analysis of the electronic mission planning system (focused on in Chapter 7) in the context of real world live exercises undertaken during an operational field trial of the digitised mission support system. DSA was analysed during mission planning and execution activities using the propositional network approach. This chapter focuses on the implications for collaborative system design that the DSA analysis and model has, in particular on the concepts of compatible and transactive SA and what these mean for collaborative system design. In closing a series of initial guidelines for collaborative system design are presented.

Chapter 9: A Model of DSA in Collaborative Systems

The final phase of the research involved developing a model of DSA in complex collaborative environments based on the findings derived from the case studies undertaken up to this point. The model of DSA is presented in Chapter 9 and its main features are discussed.

Chapter 10: Conclusions for Distributed Situation Awareness Theory, Measurement and Teamwork

The final chapter concludes the book with a discussion of the main findings from our research and the implications of the DSA model. In closing, the key contributions to knowledge brought about by our research are presented and key areas of further investigation are outlined.

Chapter 2

What Really Is Going On?
Situation Awareness Literature Review

Introduction

This chapter presents the findings derived from a literature review focusing on the area of SA in complex sociotechnical systems. The aim of the literature review was to review and critique what is currently known on SA and team SA and to identify which of the many SA models presented in the literature is currently the most suitable for describing and assessing team SA during real world scenarios undertaken in complex collaborative systems. For this purpose, a comparison of the most prominent individual and team SA models presented in the literature was undertaken. The findings derived from the literature review are discussed below.

Situation Awareness

Origins

SA is the decorative term given to the level of awareness that an individual has of a situation, an operator's dynamic understanding of 'what is going on' (Endsley, 1995a). The concept first emerged as a topic of interest within the military aviation domain when it was identified as a critical asset for military aircraft crews during the First World War (Press, 1986; cited in Endsley, 1995a). Despite this, it did not begin to receive attention in academic circles until the late 1980s (Stanton and Young, 2000), when SA-related research began to emerge within the aviation and air traffic control domains (Endsley, 1989, 1993).

Following a seminal special issue of the *Human Factors* journal on the subject in 1995, SA became a topic of considerable interest within the HF research community and many researchers began investigating it in a whole host of different domains. The construct has since evolved into a core theme within system design and evaluation, and continues to dominate HF research worldwide. For example, SA-related research is currently prominent in a wide range of areas, including the military (e.g. Stewart et al., 2008); civil aviation and air traffic control (e.g. Kaber et al., 2006); road transport (e.g. Ma and Kaber, 2007; Walker et al., 2009); energy distribution (Salmon et al., 2008b); rail (Walker et al., 2006a); naval (e.g. Stanton et al., 2009); sport (James and Patrick, 2004); health care and medicine (Hazlehurst et al., 2007) and the emergency services (e.g. Blandford and Wong, 2004). Further, a review of peer reviewed academic journal articles indicated that SA-related research has to date been reported in over 20 different scientific journals covering a diverse range of subject areas, from HF and ergonomics to sport, computer graphics, disaster response and artificial intelligence.

Defining Situation Awareness

From the onset, it is clear that the construct is a contentious one. There have been numerous attempts at defining SA and a superfluity of definitions are presented within the academic literature. It is beyond the scope of this chapter to present the range of SA definitions available in their entirety (the review identified over 30 definitions); however, a selection of the most prominent is offered below.

Based on a synthesis of fifteen SA definitions, Dominquez (1994) defined SA as an individual's, 'continuous extraction of environmental information, and integration of this information with previous knowledge to form a coherent mental picture, and the use of that picture in directing future perception and anticipating future events' (Dominguez, 1994, p. 11). Fracker (1991), defined SA as, 'the combining of new information with existing knowledge in working memory and the development of a composite picture of the situation along with projections of future status and subsequent decisions as to appropriate courses of action to take' (Fracker, 1991). Smith and Hancock (1995) describe the construct as 'externally directed consciousness' and suggest that SA is, 'the invariant in the agent-environment system that generates the momentary knowledge and behaviour required to attain the goals specified by an arbiter of performance in the environment' (Smith and Hancock, 1995, p. 145), with deviations between an individual's knowledge and the state of the environment being the variable that directs situation assessment behaviour and the subsequent acquisition of data from the environment. Bedny and Meister (1999) argue that SA is, 'the conscious dynamic reflection on the situation by an individual. It provides dynamic orientation to the situation, the opportunity to reflect not only on the past, present and future, but the potential features of the situation. The dynamic reflection contains logical-conceptual, imaginative, conscious and unconscious components which enables individuals to develop mental models of external events' (Bedny and Meister, 1999, p. 71).

The most prominent and widely used definition of SA, however, is that offered by Endsley (1995a), who defines SA as a cognitive product (resulting from a separate process labelled *situation assessment*) comprising 'the perception of the elements in the environment within a volume of time and space, the comprehension of their meaning, and the projection of their status in the near future' (Endsley, 1995a, p. 36).

Many more have attempted to define SA (e.g. Adams et al., 1995; Billings, 1995; Sarter and Woods, 1991; Taylor, 1990 etc.). Endsley's definition enjoys widespread popularity; however, many argue that a universally accepted definition of the construct is yet to emerge (e.g. Durso and Gronlund, 1999; Gorman et al., 2006; Rousseau et al., 2004; Stanton et al., 2001). It is generally agreed amongst definitions that SA refers to an individual's dynamic awareness of the ongoing external situation. The main incongruence between definitions lays in the reference to SA as either the *process* of gaining awareness (e.g. Fracker, 1991), as the *product* of awareness (e.g. Endsley, 1995a), or as a combination of the two (e.g. Smith and Hancock, 1995). This is a debate that will no doubt continue unabated; however, what is clear is that in order to fully appreciate the construct, an understanding of both the process and the product is required (Stanton et al., 2001). It is our view that, in order to produce systems that augment SA, system designers need to understand both what SA might comprise and how it is built and maintained.

Individual Models of Situation Awareness

Theoretically, the explanation of SA also remains subject to great debate. Analogous to the plethora of different SA definitions, various inconsonant theoretical perspectives have been presented and great debate still rages over which is the most appropriate account. Inaugural SA models were, in the main, focused on how individual operators develop and maintain SA whilst undertaking activity within complex systems (e.g. Adams et al., 1995; Endsley, 1995a; Smith and Hancock, 1995). Indeed, the majority of the models presented in the literature are individual focused theories, such as Endsley's three level model (Endsley, 1995a), Smith and Hancock's perceptual cycle model (Smith and Hancock, 1995) and Bedny and Meisters activity theory model (Bedny and Meister, 1999). As well as being divided by the process versus product debate, SA models also differ in terms of their underpinning psychological approach. For example, the three level model (Endsley, 1995a) is a cognitive theory that uses an *information processing* approach, Smith and Hancock's (1995) model is an ecological approach underpinned by Neisser's *perceptual cycle* model (Neisser, 1976), and Bedny and Meister's (1999) model uses an *activity theory* model to describe SA.

Various models of SA were identified via the literature review. A brief overview of the most prominent models of SA is presented below. The models considered were selected based on their description or citation in peer reviewed academic journal articles. Consequently, a small number of models presented only in book chapters (e.g. Banbury et al., 2004) and conference articles were not considered.

A number of early articles in the area attempted to clarify the cognitive underpinnings of SA using established cognitive psychology research. Sarter and Woods (1991), for example, discussed the role of various cognitive constructs in the development of SA and suggested that great care should be taken to differentiate SA from concepts such as mental models and situation assessment. They also highlighted the importance of the temporal dimension of SA, something that has been generally ignored by other researchers in the area. In discussing the cognitive processes underlying SA, Sarter and Woods (1991) suggested that SA is acquired based on the integration of knowledge that is derived from recurring situation assessments, where situation assessments are the process of perception and pattern matching (Endsley, 1988; cited in Sarter and Woods, 1991). Further, they suggested that SA refers to information that is available or that can be activated, and using Kihlstrom's (1984; cited in Sarter and Woods, 1991) taxonomy of mental contents, they argued that an individuals awareness comprises their conscious and available mental contents. Sarter and Woods (1991) subsequently defined SA as 'the accessibility of a comprehensive and coherent situation representation which is continuously being updated in accordance with the results of situation assessments' (Sarter and Woods, 1991, p. 52).

The taxonomic approach has been used by others; for instance, Taylor (1990) used interviews with military pilots to identify the dimensions associated with the process of maintaining SA. In conclusion, Taylor (1990) suggested that SA comprises three dimensions: the level of demand imposed on attentional resources by a situation; the supply of attentional resources in response to these situational demands; and the subsequent understanding of the situation. Taylor (1990) further speculated that these three dimensions contained ten generic constructs underpinning SA. These included familiarity, focusing, information quantity, instability, concentration, complexity, variability, arousal, information quality and spare capacity. Taylor (1990) subsequently described SA as 'the knowledge, cognition and anticipation of events, factors and variables affecting the safe, expedient and effective conduct of a mission' (Taylor, 1990, p. 3-3).

Endsley's three level (Endsley, 1995a) model has undoubtedly received the most attention of all of the models presented within the literature. The three level model describes SA as an internally held cognitive product comprising three hierarchical levels that is separate to the processes (termed situation assessment) used to achieve it. Endsley's model is presented in Figure 2.1.

The model depicts SA as a component of the information processing chain that follows perception and leads to decision making and action execution. According to the model, SA acquisition and maintenance is influenced by various factors including individual factors (e.g. experience, training, workload etc.), task factors (e.g. complexity) and systemic factors (e.g. interface design) (Endsley, 1995a).

Endsley's account focuses on the individual as a passive information receptor and divides SA into three hierarchical levels. The first step involves perceiving the status, attributes and dynamics of task-related elements in the surrounding environment (Endsley, 1995a). At this stage, the data is merely perceived and no further processing of the data takes place. The data perceived is dependent on a range of factors, including the task being performed, the operator's goals, experience, expectations and systemic factors such as system capability, interface design, level of complexity and automation.

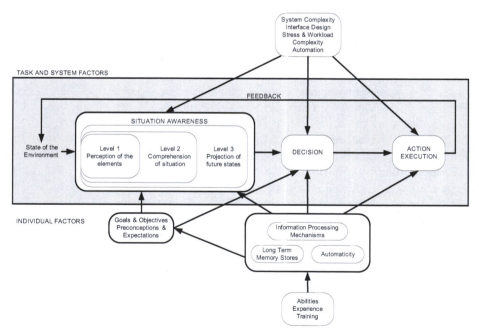

Figure 2.1   The three level model of situation awareness

Source: Adapted from Endsley, 1995a.

According to Endsley (1995a), 'a person's goals and plans direct which aspects of the environment are attended to in the development of SA' (Endsley, 1995a, p. 47).

Level 2 SA involves the interpretation of level 1 data in a way that allows an individual to comprehend or understand its relevance in relation to their task and goals. During the acquisition of level 2 SA 'the decision maker forms a holistic picture of the environment, comprehending the significance of objects and events' (Endsley, 1995a, p. 37). Similarly to level 1 SA, the interpretation and comprehension of SA-related data is influenced by an individual's goals, expectations, experience in the form of mental models, and preconceptions regarding the situation. Key here is the use of experience in the form of mental models to facilitate the acquisition of level 2 SA. Endsley suggests that more experienced operators use mental models to facilitate this integration of level 1 elements and goals to achieve comprehension.

The highest level of SA involves prognosticating the future states of the system and elements in the environment. Using a combination of level 1 and 2 SA-related knowledge, and experience in the form of mental models, individuals can forecast likely future states in the situation. For example, a military pilot forecasts, based on level 1 and level 2-related information (e.g. location, positioning, objectives etc.) and experience, that an aircraft might attack in a certain manner (Endsley, 1995a). The pilot can do this through perceiving and understanding the speed, location, formation and movements of enemy aircraft and comparing this to experience (in the form of mental models) of similar situations. This comparison of situational data with past experience allows operators to project future situational states.

One of the key assumptions of the three level model is the critical role of mental models in the development and maintenance of SA. According to Endsley (1995a), features in the environment are mapped to mental models in the operators mind, and the models are then used to facilitate the development of SA. Mental models (formed by training and experience) are used to facilitate the achievement of SA by directing attention to critical elements in the environment (level 1), integrating the elements to aid understanding of their meaning (level 2) and generating possible future states and events (level 3).

Bedny and Meister (1999) describe SA through a theory of activity approach that outlines the various cognitive processes that are associated with human behaviour. The theory of activity itself purports that individuals possess goals, which represent an ideal image or desired end state of activity, motives that direct them towards the end state, and methods of activity (or actions) that permit the achievement of these goals (Bedny and Meister, 1999). Differences between the goals and the current situation motivate an individual to take action towards achieving the goal. According to the theory, activity comprises three stages: the orientational stage, the executive stage and the evaluative stage (Bedny and Meister, 1999). The orientational stage involves the development of an internal representation or picture of the world or current situation. The executive stage involves proceeding towards a desired goal via decision-making and action execution. Finally, the evaluative stage involves assessing the situation via information feedback, which in turn influences the executive and orientational components. The functional model of activity is presented in Figure 2.2.

Bedny and Meister (1999) suggest that each of the functional blocks presented in Figure 2.2 has a specific role to play in the development and maintenance of SA and that the blocks orientate themselves towards the achievement of SA. The interpretation of

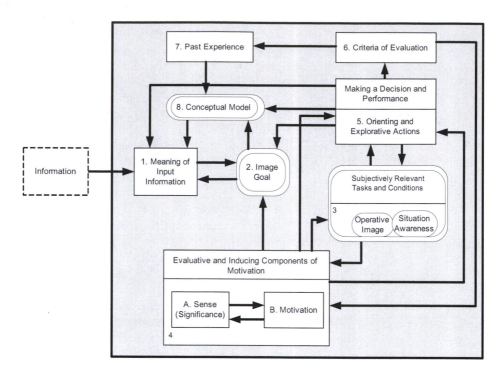

Figure 2.2   Interactive subsystems approach to situation awareness

Source: Adapted from Bedny and Meister 1999.

incoming information (function block 1) is influenced by an individual's goals (function block 2), conceptual model of the current situation (function block 8) and past experience (function block 7). This interpretation then modifies an individual's goals, experience, and conceptual model of the current situation. Critical environmental features are then identified (function block 3) based on their significance to the task goals and the individual's motivation towards the task goal (function block 4), which directs their interaction with the world (function block 5). The extent to which the individual proceeds to engage the task goals is determined by their goals (function block 2) and their evaluation of the current situation (function block 6). The resultant experience derived from the individual's interaction with the world is stored as experience (function block 7), which in turn informs their conceptual model (function block 8). According to the model, the core processes involved in the acquisition of SA are the conceptual model (functional block 8), the image-goal (functional block 2) and the subjectively relevant task conditions (functional block 3).

Taking the example of a military commander, Bedny and Meister's model suggests that commanders interpret information received from the world (function block 1) based on their past experience and knowledge of similar situations (function block 7), their abstract model of the world (function block 8) and mission goals (function block 2). The commanders then identify task critical cues in the environment (function block 3) based upon their significance to their goals, their motivation towards these goals (function block

4) and their interaction with the world (function block 5). Based on the comprehension of these task critical cues, their goals and motivation, the commanders then decides on an appropriate course of action (function block 5) and interacts with the world accordingly (i.e. communicates with subordinates, monitors outcomes and receives feedback). Their interactions and outcomes are then stored as experience (function block 7), which in turn informs their representation or conceptual model of the world (function block 8).

Smith and Hancock's (1995) ecological approach takes a more holistic stance, viewing SA as a 'generative process of knowledge creation and informed action taking' (1995, p. 138). Their description is based uponon Neisser's (1976) perceptual cycle model, which describes an individual's interaction with the world and the influential role of schemata in these interactions. According to the perceptual cycle model our interaction with the world (termed explorations) is directed by internally held schemata. The outcome of interaction modifies the original schemata, which in turn directs further exploration. This process of directed interaction and modification continues in an infinite cyclical nature.

Using this model, Smith and Hancock (1995) suggest that SA is neither resident in the world nor in the person, but resides through the interaction of the person with the world. Smith and Hancock (1995, p. 138) describe SA as 'externally, directed consciousness' that is an 'invariant component in an adaptive cycle of knowledge, action and information'. Smith and Hancock (1995) argue that the process of achieving and maintaining SA revolves around internally held mental models, which contain information regarding certain situations. These mental models facilitate the anticipation of situational events, directing an individual's attention to cues in the environment and directing their eventual course of action. An individual then conducts checks to confirm that the evolving situation conforms to their expectations. Any unexpected events serve to prompt further search and explanation, which in turn modifies the operators existing model. The perceptual cycle model of SA is presented in Figure 2.3.

Unlike the three level model (and similar to the activity theory model), which depicts SA as a product separate from the processes used to achieve it, SA is viewed a both process and product, offering an explanation of the cognitive activity involved in achieving SA and also a judgement as to what the product of SA comprises. Smith and Hancock's (1995) complete model therefore views SA as more of a holistic process that influences the generation of situational representations. For example, in reference to air traffic controllers 'losing the picture', Smith and Hancock suggest, 'SA is not the controller's picture. Rather it is the controllers SA that builds the picture and that enables them to know that what they know is insufficient for the increasing demands.' (Smith and Hancock, 1995, p. 142).

In a similar fashion, Adams et al. (1995) used a modified version of Neisser's (1976) perceptual cycle model in an attempt to clarify the cognitive components involved in the acquisition and maintenance of SA. Adams et al. (1995) mused on the interdependence between the process of SA and the product of SA and subsequently argued that Neisser's (1976) model could be used to describe how SA is acquired and maintained, suggesting that 'SA can be seen as both product and process. As product, it is the state of the active schema-the conceptual frame or context that governs the selection and interpretation of events. As process, it is the state of the perceptual cycle at any given moment. As process and product, it is the cyclical resetting of each by the other' (Adams et al., 1995, p. 89). Further, they used Sanford and Garrod's (1981; cited in Adams et al., 1995) work

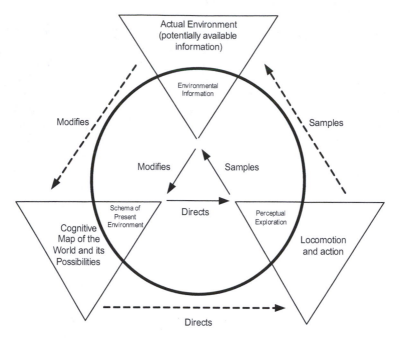

Figure 2.3   The perceptual cycle model of SA

Source: Adapted from Smith and Hancock, 1995.

on explicit and implicit focus in working memory to speculate on the mechanisms of SA. This suggests that working memory contains two bins, explicit focus and implicit focus, and that long-term memory too contains two bins, episodic memory and semantic memory. Adams et al. subsequently added these aspects to the perceptual cycle model in order to explain how SA is achieved and maintained. According to Adams et al., the explicit and implicit focus bins could replace the 'schema of present environment' box within the perceptual cycle (see Figure 2.3) and the episodic and semantic memory bins could replace the 'cognitive map of the world and its possibilities' box.

More recently, Hourizi and Johnson (2003) developed and tested a model of awareness based on existing cognitive constructs. Inspired by interacting cognitive sub-systems theory, they suggest that information has to pass through a series of cognitive processes before it can be considered as 'awareness' (Hourizi and Johnson, 2003). Hourizi and Johnson's model suggests that awareness involves information in the world passing through four sub-processes: 1) information is available in the world; 2) information is perceived; 3) information is attended to; and 4) information is subject to higher level cognitive processing. For example, their account suggests that, within the cockpit, information that is available in the world is perceived by the pilot. This information is then attended to by the pilot, following which it is subject to further, higher level semantic cognitive processing which allows the pilot to understand the implication of the information in question.

Similarities between Hourizi and Johnson's model and Endsley's three level model are apparent. Level 2 in this case is similar to Endsley's level 1, where information in the world is perceived but no interpretation occurs; Level 4 in Hourizi and Johnson's model ostensibly is similar to Endsley's level 2, where the information perceived is understood in light of ones goals.

Hourizi and Johnson (2003) suggest that the model can be used to explain different SA breakdowns within the cockpit by specifying the SA sub-process that failed. They cite the example of information being available in the cockpit but not being seen by the pilot as a failure of level 2 (perception), the instance where information is picked up visually but is not attended to due to the pilots attention being elsewhere as an example of a level 3 failure, and the instance where information is available, seen, attended to, but not understood as a level 4 failure. They suggest that the model can be used to ask focussed questions (i.e. at each level) during the early stages of the design process in order to predict potential SA-related problems.

Summary of Individual SA Models

In order to identify the most appropriate individual SA model for describing SA in complex sociotechnical systems, the different models described above were evaluated. For this purpose, a set of criteria for describing and evaluating the models was developed based on existing HF review articles and also other HF theory and methods criteria taken from the literature (e.g. Kirwan, 1992, 1998; Salmon et al., 2006; Stanton and Young, 1999b; Stanton et al., 2004; Stanton et al., 2005). The criteria used included:

1. *Model name and acronym* – presents the name of the model and any associated acronyms;
2. *Domain of origin* – details the domain in which the model originally emerged;
3. *Domain(s) of application* – details the domains in which the model has subsequently been applied;
4. *Theoretical underpinning* – describes the use of any established psychological theory to underpin the model;
5. *Process* – summarises the process of developing and maintaining SA according to the model;
6. *Composition* – describes the composition of SA according to the model;
7. *Novelty* – describes the novelty of the model based on how it differs from existing psychological models;
8. *Measure* – describes any related SA measurement approaches;
9. *Process or Product* – delineates whether the model describes SA as a process, a product or as a combination of the two;
10. *Citation* – depicts the number of citations derived from an analysis of peer-reviewed journal articles (as of January 2008);
11. *Main strengths* – lists the main strengths of the model in relation to its use in system design and evaluation; and
12. *Main weaknesses* – lists the main weaknesses of the model in relation to its use in system design and evaluation.

The individual SA models are evaluated in Table 2.1.

Table 2.1   Individual SA theory comparison table

Theory	Domain of Origin	Domain Applications	Theoretical Underpinning	Process	Composition	Novelty
Three Level Model (Endsley, 1995a)	Aviation	Military (Aviation, Infantry Ops), Air Traffic Control, Aviation (Flight & Maintenance), Driving, Nuclear Power,	Information Processing Theory Recognition Primed Decision Making Model (Klein, 1990)	Perception of elements Comprehension of meaning Projection of future states	Perception, comprehension and projection of SA elements	Not distinct from information processing models
Perceptual Cycle Model (Smith & Hancock, 1995)	Air Traffic Control	Air Traffic Control	Perceptual Cycle Model (Niesser, 1976)	Schema driven exploration & modification	Externally directed consciousness	Subset of activated schema which is externally focussed
Theory of Activity Model (Bedny & Meister, 1999)	N/A	N/A	Theory of Activity Model (Bedny & Meister, 1999)	Orientational Stage Executive Stage Evaluative Stage	Incoming Information Goals Conceptual Model of Situation Past Experience Environmental Features Motivation towards task goals Subjectively relevant Task Conditions	Separate functional block within activity theory model
Sarter & Woods (1991)	Aviation	Aviation	Working Memory Mental Models Situation Assessment Awareness	Integration of knowledge derived from situation assessments	Accessible and Activated Knowledge Conscious and available mental contents	Activated knowledge in working memory derived from situation assessment
Adams, Tenney & Pew (1995)	Aviation	Aviation	Perceptual Cycle Model (Niesser, 1976) Working Memory Theory (Sanford & Garrod, 1981)	Schema driven exploration & modification	Active Schema Explicit and Implicit focus in Working Memory	Activated Schema
Predictive account of awareness Hourizi & Johnson (2003)	Aviation	Aviation	Interactive Cognitive Sub-systems (Bernard & May, 2000) Rushby (1999)	1) information availability ; 2) perception of information ; 3) attention ; and 4) Higher level cognitive processing	Perceptual level awareness Semantically Processed level of awareness	Differentiation between perception and semantic level SA
Taylor (1990)	Aviation	Military (Aviation, Infantry Ops), Air Traffic Control, Aviation (Flight & Maintenance), Driving, Nuclear Power,	Theories of Attention and Cognition	1. Attentional Demand 2. Supply of Attentional Resources 3. Situational Understanding	Demand, Supply of attentional resources , Understanding 10 Dimensions - familiarity, focusing, information quantity , instability, concentration , complexity, variability, arousal , information quality and spare capacity.	Multi-dimensional charcterisation of SA

Endsley's three level model is undoubtedly the most popular and widely applied of the models described with over 50 citations in the peer reviewed HF literature. On a positive note, the model is generic and presents a simplistic and intuitive description of SA, which has subsequently led to its application in a plethora of different domains. Further, the popularity of the model is such that it has been extended in order to describe team SA (e.g. Endsley and Jones, 2001; Endsley and Robertson, 2000). The model's utility lies in its simplicity and also the division of SA into three hierarchical levels, which allows the construct to be measured easily and effectively and also supports the abstraction of SA requirements (e.g. Matthews et al., 2004) and the development

Table 2.1   Individual SA theory comparison table

Measure	Process or Product?	Citations	Main Strengths	Main Weaknesses
SAGAT (Endsley, 1995b) SA Requirements Analysis (Endsley, 1993)	Product	52	1. Simple intuitive description of SA 2. Division of SA into levels is neat and permits measurement using SAGAT approach 3. Holistic approach that considers factors such as system & interface design, workload and training	1. Fails to cater for the dynamic nature of SA 2. SA process oriented definition is contradictory to the description of SA as a 'product' comprising three levels 3. Based on ill defined and poorly understood psychological models (e.g. information processing, mental models)
Task Performance Risk Space (Smith & Hancock, 1995)	Process & Product	5	1. Dynamic description of SA acquisition, maintenance and update of schema 2. Sound theoretical underpinning 3. Completeness of model is attractive i.e. it describes both the process of acquiring SA and the product of SA	1. Does not translate easily to SA description and measurement 2. Limited applications 3. The actual correlation between SA and performance is complex and not yet fully understood
N/A	Process & Product	4	1. Model offers a more complete, dynamic description of SA than the three-level model 2. Clear description of each functional blocks role in SA acquisition and maintenance is useful 3. SA described as a distinct and separate entity	1. Limited application and the model lacks supporting empirical evidence 2. Underpinning activity theory remains unclear 3. No measurement approach suggested
Embedded Tasks	Process & Product	16	1. Explicitly considers the temporal dimensions of SA 2. Attempts to differentiate between SA and existing psychological constructs such as mental models and awareness	1. More of a discussion than a model 2. Model not since advanced through study 3. Embedded tasks measurement approach has received only limited attention
Computational Modelling	Process & Product	9	1. Sound theoretical underpinning 2. Dynamic description of SA acquisition, maintenance and update of schema 3. Completeness of model is attractive i.e. it describes both the process of acquiring SA and the product of SA	1. The use of computational modelling to measure SA is questionable 2. Limited applications of the model; theory not since advanced
Undesirable Interventions Man-Machine Interactions Verbal Protocol Analysis Participant De-Brief	Process & Product	1	1. Attempts to differentiate between perceptual level of SA and semantically processed level of SA 2. Use theory to test interface design concept	1. Limited explanation of the process of acquiring and maintaining SA 2. Measurement approach requires simulation 3. Limited applications within published literature
Situation Awareness Rating Technique (Taylor, 1990)	Process & Product	7	1. Attempts to identify the different dimensions comprising SA 2. Proposes measurement approach based on SA dimensions 3. Based on pilot knowledge elicitation exercise	1. No description of the process of acquiring and maintaining SA 2. Measurement approach (SART) has performed poorly in validation studies 3. Does not clearly explain link between workload and SA

of training strategies and design guidelines (e.g. Endsley et al., 2003) to support the acquisition of the different SA levels. Endsley's model is also comprehensive in that it identifies the various different factors (individual, task and system) that affect an individual's acquisition and maintenance of SA. The notion that experienced operator's use internally held mental models formed by experience in order to facilitate the development of the higher levels of SA is also fitting, and can be used to effectively explain the differences between the levels of SA achieved by novices and experts. This concept is also underpinned by Klein's theory of recognition-primed decision-making (Klein, 1998).

Despite its popularity, the three level model is not without its flaws and has been subject to various criticisms. The model is ostensibly a linear feedback model of SA and so ignores the notion that SA can be as much a feed forward phenomena as a feedback one. For example, driving without attention mode (Kerr, 1991; May and Gale, 1998) describes how drivers arrive at destinations without being aware of how they got there and relates to the notion that experts are likely to be able to generate SA without necessarily perceiving all of the elements in the environment (i.e. project with perception etc.). Such a driver would, of course, be deemed to have poor SA under Endsley's model. In addition, many have questioned the similarities between Endsley's model and the construct of working memory, which in turn has led to them questioning the notion that SA represents a construct in its own right. There is also a lack of empirical evidence supporting the model and it is questionable whether or not a testable hypothesis could in fact be generated using this perspective (although this is perhaps a criticism that can be levelled at all SA models).

Taking a closer look at the model reveals other problems. The description of SA being ones awareness, comprehension and projection of 'elements' in the environment is all very well, however, there is little consideration given to the links and interactions between these elements and the individuals cognisance of this. The linkage between elements could conceivably determine the character of SA as much as the elements themselves. Also, Endsley distinguishes between the product of SA and the processes that are used to achieve it and suggests that the two are separate. Her account is therefore contradictory, since it refers to the 'perception of the elements', the 'understanding of their meaning' and the 'projection of future states', all of which could be taken to be processes involved in the development of SA. The model can also be criticised for its sequential (or linear) description of the process of achieving SA (i.e. level 1 then level 2 then level 3) since it may be that higher levels of SA, particularly level 3 can be achieved without the development of the preceding levels.

A number of researchers have also criticised the model for basing its theoretical foundations on what are evidently poorly understood constructs themselves. Smith and Hancock (1995), for example, suggest that its reference to mental models, which themselves are ill defined, is problematic. Similarly, Uhlarik and Comerford (2002) criticise Endsley's model for its use of an information-processing model containing psychological constructs that are not yet fully understood and that are subject to great debate themselves. The model has also been criticised for its inability to cope with the dynamic nature of SA (Stanton et al., 2001). Uhlarik and Comerford (2002), for example, point out that the process of achieving SA presented by the three level model is both static and finite.

The activity theory model description presented by Bedny and Meister offers perhaps a more dynamic description of the process of acquiring SA. In particular, the description of the way in which SA dynamically modifies interaction with the world and then interaction with the world dynamically modifies SA is logical, and goes beyond the static perspective taken by Endsley's model. Moreover, the clear elucidation of each of the functional blocks roles in the development of SA is useful.

The activity model, however, is also not without its flaws. Activity theory itself has not yet been fully embraced and there is a distinct lack of empirical evidence that supports the model. The model has also received far less attention than the three level and perceptual cycle models. Further, the process and product approach adopted by the model makes the measurement of SA from this perspective very difficult. In addition, like most other SA related models, the activity theory model does not attempt to cater for, or explain, team or shared SA and has not been extended to do so. Finally, the model lacks ecological validity since it only shows a one-way link between function blocks 2 and 4. Also there is no link to the world from function block 5; the model is therefore a closed loop since there is no output or feedback prescribed within it.

The perceptual cycle model (Smith and Hancock, 1995) offers a complete description of how SA is achieved and maintained. The model has sound underpinning theory (Neisser, 1976) and is complete in that it refers to the continuous cycle of SA acquisition and maintenance, including both the process (the continuous sampling of the environment) and the product (the continually updated schema) of SA. Their description also caters for the dynamic nature of SA and more clearly describes an individual's interaction with the world in order to achieve and maintain SA whereas Endsley's model seems to place the individual as a passive information receiver. The model therefore considers the individual, the situation and the interactions between the two. The definition and model presented by Smith and Hancock (1995), in the authors' opinion at least, comes closest to most accurately describing the construct. Adams et al.'s (1995) perceptual cycle model is also attractive for the same reasons, but mainly for its ability to describe how SA is dynamically acquired, maintained and updated. Adams et al. (1995) also logically use the model to explain anticipation (or level 3 SA as described by Endsley).

Disappointingly, however, neither Adams et al. (1995) nor Smith and Hancock's models have received anywhere near the attention that Endsley's model has. Further, measuring SA in line with the perceptual cycle models is difficult since they consider internally held schema and externally directed consciousness, both of which are difficult to assess. It is also notable that, despite their attractiveness as accounts of SA, the perceptual cycle models have not yet been extended in order to describe team SA.

Aside from Endsley's model, all of the other approaches described have received only very limited attention. Sarter and Woods' (1991) approach is useful in that it focuses on the temporal dimensions of SA and emphasises the differences between SA, mental models and situation assessment. Taylor's (1990) work has received attention but more so for the Situation Awareness Rating Technique (SART; Taylor, 1990) measurement approach that was subsequently developed.

To summarise, each theory has its useful components. Stanton et al. (2001) point out that three level model, the perceptual cycle model and the activity theory model all have an element of truth in them. The definition of SA as externally directed consciousness, in our opinion, certainly holds the most credibility and in terms of theoretical utility. Smith and Hancock's and Adams et al.'s models are perhaps the most useful since they cater for the dynamic aspects of SA and also for both the processes involved and the end product SA. Endsley's three level model, on the other hand, offers an intuitive description of SA which allows researchers to measure it simplistically and also to abstract SA requirements at each level. The description of the individual, task and system factors affecting SA acquisition and maintenance offered by Endsley (1995a) is also useful. It is perhaps for these reasons that it has been embraced by researchers wishing to describe and measure the nature and content of operator SA in complex systems.

Disappointingly, there is no empirical evidence that directly validates the models discussed. If SA is a cognitive phenomenon it cannot be observed directly, which makes the validation of SA models somewhat difficult. Further, the extent to which SA represents a psychological construct in its own right has been questioned. Moray (2004) for example defines SA not as a unique psychological function, but simply as the ability to 'keep track of what is going on around you in a complex and dynamic environment' (p. 4). Others have pointed out strong similarities with the construct of working memory. Bell and Lyon, for example, suggest that SA is 'eventually reducible to some form of ... information in working memory' (Bell and Lyon, 2000, p. 42). It is worth pointing out here that this is not the case where SA is viewed as a 'social' phenomenon (e.g. Hutchins, 1995); in this case, SA is taken to reside in the artefacts and conversations around us.

Aside from judgements on the validity of the different models discussed, the overarching conclusion to take from the individual SA models review is that none of the models, in their present form at least, are easily extendable to the description of SA within collaborative environments; all focus exclusively on SA 'in the head' of individual operators and so cannot be used to describe SA during collaborative endeavour. The next section of this review therefore considers the different team SA models presented in the literature.

Situation Awareness in Collaborative Systems

Teams

The use of teams has increased significantly over the past three decades (Savoie, cited in Salas, 2004). This is primarily due to two factors; firstly the increasing complexity of work and work procedures and secondly because appropriately trained and constructed teams can potentially offer a number of advantages over the use of individual operators, including the ability to perform more difficult and complex tasks better, greater productivity and improved decision making (Orasanu and Fischer, 1997), more efficient performance under stress (Salas and Cannon-Bowers, 2004) and a reduction in the number of errors made (Wiener et al., 1993; cited in Salas and Cannon-Bowers, 2004).

Due to the significant presence of teams in contemporary systems, and the likelihood that their presence will increase significantly in the future (Fiore et al., 2003), the construct of team SA is currently receiving increased attention from the HF community. A team is characterised as consisting of two or more people, dealing with multiple information sources and working to accomplish a shared or common goal of some sort. Salas et al. (1995, p. 127) define a team as 'a distinguishable set of two or more people who interact dynamically, interdependently and adaptively toward a common and valued goal, who have each been assigned specific roles or functions to perform and who have a limited life span of membership'.

Collaborative endeavour comprises two forms of activity: teamwork and taskwork. Teamwork refers to those instances where individuals interact or coordinate behaviour in order to achieve tasks that are important to the team's goals (i.e., behavioural, attitudinal and cognitive responses coordinated with fellow team members), whilst taskwork (i.e., task-oriented skills) describes those instances where team members are performing individual tasks separately from their team counterparts. Wilson et al. (2007) define teamwork as 'a multidimensional, dynamic construct that refers to a set of interrelated cognitions, behaviours and attitudes that occur as team members perform a task that results in a coordinated and synchronised collective action.' According to Glickman et al. (1987; cited in Burke, 2004), team tasks require a combination of taskwork and teamwork skills in order to be completed effectively.

Team Situation Awareness

Team SA is indubitably more complex than individual SA. Salas, Prince, Baker and Shrestha (1995) point out that there is a lot more to team SA than merely combining individual team member SA. Further, Salas, Muniz and Prince (2006) argue that, due to the cognitive nature of team SA, research into the construct is difficult, deficient and complex. Consequently, team SA suffers from a similar level of contention as the area of individual SA does.

Ostensibly team SA is multi-dimensional, comprising individual team member SA, shared SA between team members and also the combined SA of the whole team, the so-called 'common picture'. Add to this the various team processes involved (e.g. communication, coordination, collaboration, etc.) and the complexity of the construct quickly becomes apparent. Most attempts to understand team SA have centred on a 'shared understanding' of the same situation. Nofi (2000, p. 12), for example, defines team SA as 'a shared awareness of a particular situation' and Perla et al. (2000, p. 17) suggest that 'when used in the sense of 'shared awareness of a situation', shared SA implies that we all understand a given situation in the same way'. Stout et al. (cited in Salas et al., 2006) suggested that team SA comprises each team members SA and the degree of shared understanding between team members.

Based on a review of the literature, Salas et al. (1995) proposed a framework of team SA, suggesting that it comprises two critical, but poorly understood, processes individual SA and team processes. According to Salas et al. (1995), team SA depends on communications at various levels. The perception of SA elements is influenced by the communication of mission objectives, individual tasks and roles, team capability and other team performance factors. Salas et al. (1995) suggested that schema limitations can be offset by information exchange and communication, the information to support this being provided by communication and coordination between team members. The comprehension of this information is affected by the interpretations made by other team members, so it is evident that SA leads to SA and also modifies SA, in that individual SA is developed, and then shared with other team members, which then develops and modifies team member SA. Thus, a cyclical nature of developing individual SA, sharing SA with other team members and then modifying SA based on other team members SA is apparent. Salas et al. (1995) also highlighted the importance of team processes such as communication, assertiveness and planning, all of which they suggest contribute to the acquisition and maintenance of team SA. Salas et al. (1995) subsequently define team SA as 'the shared understanding of a situation among team members at one point in time (Salas et al., 1995, p. 131) and concluded that team SA 'occurs as a consequence of an interaction of an individual's pre-existing relevant knowledge and expectations; the information available from the environment; and cognitive processing skills that include attention allocation, perception, data extraction, comprehension and projection' (Salas et al., p. 125). Salas et al.'s model of team SA is presented in Figure 2.4.

Wellens (1993) used a model of distributed decision-making (Wellens and Ergener, 1988; cited in Wellens, 1993) to describe SA during collaborative activity. Wellens (1993) suggested that the key to team SA lies in the arrangement of teams so that sufficient overlap between team member SA occurs to support coordination, but also so that sufficient separation between members allows individual SA acquisition. Wellens (1989a; cited in Wellens, 1993, p. 272) defined group or team SA as 'the sharing of a common perspective between two or more individuals regarding current environmental events, their meaning and projected future'.

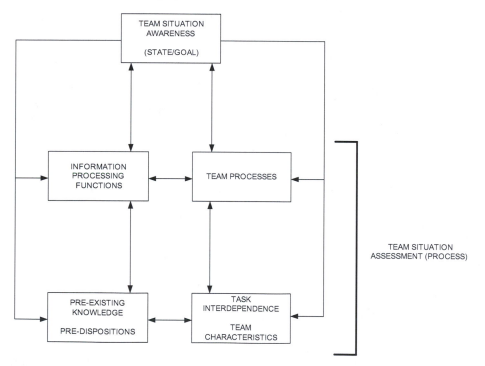

Figure 2.4   Team situation awareness model

Source: Adapted from Salas et al., 1995.

Shu and Furuta (2005) recently proposed a novel theory of team SA based on Endsley's (1995a) model and Bratman's (1992; cited in Shu and Furuta, 2005) theory of shared cooperative activity. They suggested that team SA comprises both individual SA and mutual awareness (the awareness that a cooperative entirety have of each other's activities, beliefs and intentions) and can be described as a partly shared and partly distributed understanding of situation among team members. Shu and Furuta (2005) defined team SA as, 'two or more individuals share the common environment, up-to-the-moment understanding of situation of the environment, and another person's interaction with the cooperative task.' (Shu and Furuta, 2005, p. 274).

Shared Situation Awareness

Endsley (1995a) and Endsley and Jones (1997) make the distinction between team SA and shared SA. Shared SA refers to the level of overlap in common SA elements between team members. That is, each team member has specific SA requirements for their task, some of which may overlap with other team members' requirements. Shared SA is defined as 'the degree to which team members have the same SA on shared SA requirements' (Endsley and Jones, 1997, p. 54). Team SA, on the other hand, is defined as 'the degree to which every team member possesses the SA required for his or her responsibilities' (Endsley, 1995b, p. 31). Endsley (1995b) suggests that, during team activities, SA can overlap between team members, in that individuals need to perceive, comprehend and project SA elements that are specifically related to their specific role in the team, but also elements that are required by themselves and by other members of the team. This is represented in Figure 2.5.

■ = Individual team member SA elements

◯ = Shared SA elements

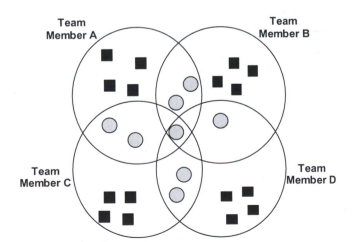

Figure 2.5   Team and shared situation awareness

Source: Adapted from Endsley, 1995b.

Endsley's shared SA account has been applied in a number of domains. For example, in the conclusion to a review of team SA in aircraft maintenance teams, Endsley and Robertson (2000) suggested that good team SA is dependent on team members understanding the meaning of the information that is passed between them. According to Endsley and Robertson (2000), this means that teams need to share pertinent data and the higher levels of SA, such as the significance of SA elements to the team's goals and projected states. Endsley and Robertson (2000) go on to suggest that the primary factors linked to team performance are shared goals, the interdependence of team member actions and the division of labour between team members. This means that some SA requirements are independent but also that team members possess shared goals and perform interdependent activities, which means that they also possess shared SA requirements. Efficient team performance, according to Endsley and Robertson, is dependent on team members having good SA on their own SA requirements and the same SA on shared SA requirements.

Development of Team and Shared Situation Awareness

The role of team behaviours, such as coordination, collaboration, adaptability and cooperation (Fiore et al., 2003) and team attitudes, such as team trust and cohesion, collective efficacy and orientation (Fiore et al., 2003), is often only briefly considered when describing team SA. Salas et al. (1995), for example, point out that there has been little consideration of the effects of team process variables on team SA. It seems logical to assume that an increased level of teamwork will lead to enhanced levels of team SA; however, the specific relationship between team behaviours and attributes and team SA remains largely unexplained. Most researchers have focused on communication as the key element in the acquisition of team SA. Nofi (2000), for example, cites communication as the most critical element in the creation of team or shared SA and Entin and Entin (2000) report that communication is a prerequisite for high levels of team SA. Salas et al. (1995) suggest that those team processes that facilitate communication, such as assertiveness, planning and leadership, contribute to team SA development. Salas, Burke and Samman (2001) suggest that one of the key factors in facilitating shared SA is a climate that supports clear and open communication. Endsley (1995a) mirrors this view by suggesting that the team member SA of shared elements may provide an index of teamwork (i.e. coordination) or team communications.

Bolstad and Endsley (2000) propose that the development of shared SA involves the following four factors: shared SA requirements (e.g. the degree to which team members understand which information is needed by other team members), shared SA devices (e.g. communications, shared displays and the shared environment), shared SA mechanisms (e.g. shared mental models) and shared SA processes (effective team processes for sharing relevant information). Lloyd and Alston (2003) argue that team members acquire individual SA and then communicate this throughout the team, which leads to a common team understanding. Another key aspect of teamwork that is critical to team SA is the process of mutual monitoring, whereby team members monitor one another's activities (e.g. Rognin et al., 1998), allowing the extraction of situational information without explicit verbal communication and also other team members' understanding of it. Mutual performance monitoring represents the 'the ability to keep track of fellow team members work, while carrying out their own work, to ensure that everything is running as expected and to ensure that they are following procedures correctly' (Wilson et al., 2007); this requires that team members have an understanding of the individual team members and overall team tasks, as well as an awareness of the team members roles, responsibilities and an expectation of what team members should be doing.

Another key concept thought to be critical to team SA is the notion of shared mental models. Mental models are essentially internal representations of a system or process and have been defined as 'knowledge structures, cognitive representations or mechanisms which humans use to organise new information, to describe, explain and predict events as well as to guide their interactions with others' (Paris et al., 2000). Fiore et al. (2003) suggest that a shared mental model is, 'the activation in working memory of team and task-related knowledge while engaged in team interaction'. According to Klein (2000), shared mental models refer to the extent that members have the same understanding of the dynamics of key processes; for example, roles and functions of each team member, nature of the task and use of equipment. Stout et al. (1999) suggest that shared mental models 'are thought to provide team members with a common understanding of who is responsible for what task and what the information requirements are. In turn, this allows them to anticipate one another's needs so that they can work in sync'. According to Cannon-Bowers et al. (1993; cited in Salas et al., 1995), shared mental models are organised bodies of knowledge that are shared across members of a team (Cannon-Bowers et al., 1993; cited in Salas et al., 1995). Cannon-Bowers and Salas (1997) suggest that shared mental models contain overall task and team goals and knowledge of individual tasks and team member roles. Endsley and Jones (1997) argue that shared mental models should incorporate an understanding of other teams' roles, plans, information requirements and potential new re-plans, and the ability to project the actions and responses of other teams.

The importance of shared mental models in the development and maintenance of team SA has been postulated by a number of researchers in the field. According to Fox et al. (2000), effective team functioning requires the existence of a shared or team mental model among members of a team. Fiore et al. (2003) suggest that effective teams develop shared mental models that they use to coordinate behaviour. It is also thought that shared mental models facilitate communications between team members (Perla et al., 2000) and can allow team members to forecast the behaviour of other team members (Fiore et al., 2003; Salas et al., 1994). Cannon-Bowers et al. (1993; cited in Salas et al., 1995) suggest that when communications channels are limited, shared mental models allow team members to anticipate other team member behaviours and information requirements. Further, they suggest that shared mental models of team tasks allow team members to perform functions from a common frame of reference. Endsley (1995) argues that team SA is more reliant on shared mental models than it is on verbal communication.

Distributed Situation Awareness

A more recent theme to emerge within the SA literature is the concept of distributed or systemic SA. Distributed SA (DSA) approaches were born out of distributed cognition theory (Hutchins, 1995), which describes the notion of joint cognitive systems comprising the operators working within the system and the artefacts that they use. According to distributed cognition theory, cognition transcends the boundaries of individual actors and instead becomes a systemic function that is achieved by coordination between the human and technological agents working within the collaborative system; cognition is achieved through coordination between system units (Artman and Garbis, 1998) and is therefore viewed as an emergent property (i.e. relationship between systemic elements) of the system rather than an individual endeavour. Inspired by distributed cognition theory, DSA approaches thus use the system itself as the unit of analysis when studying SA rather than the individuals within it. This is in line with Ottino's (2003) assertion that, 'complex systems cannot be understood by studying parts in isolation. The very essence of the system lies in the interaction between parts and the overall behaviour that emerges from the interactions. The system must be analysed as a whole' (p. 293).

DSA approaches therefore view team SA not as a shared understanding of the situation, but rather as an entity that is separate from team members and is in fact a characteristic of the system itself (Artman and Garbis, 1998). This contradicts those models that view SA as a uniquely cognitive construct (e.g. Endsley, 1995a; Sarter and Woods, 1991) and instead takes a systems view on SA. Whilst recognising that individuals within a team possess their own SA for a particular situation and that team members may share their understanding of the situation (Artman and Garbis, 1998), DSA approaches assume that collaborative systems posses cognitive properties (such as SA) that are higher than individual cognition. SA, as are other systems level cognitive processes, is therefore taken to be an emergent property of collaborative systems. It resides in the cycle of activity rather than any one agent alone and may be associated with agents but does not reside within them as it is born out of the interactions between them.

Artman and Garbis (1998) were the proponents of the distributed cognition approach to team SA, suggesting that when considering team performance in complex systems, it is necessary to focus on the joint cognitive system as a whole. In their description of SA as distributed cognition, they suggest that the SA of a team is distributed not only throughout the agents comprising the team, but also in the artefacts that they use in order to accomplish their goals. They also argued that in complex sociotechnical domains such as military command, a non-individual approach to the assessment of SA is necessary. Their distributed cognition approach focuses on the interactions between team members and artefacts rather than mental process and takes the joint cognitive system as the unit of analyses rather than the individual (Artman and Garbis, 1998). In looking at the construct of SA in this manner, it is assumed that the teams awareness of the situation is distributed throughout the joint system comprised of team members and the artefacts that they are using. No one member has the overall SA; rather it is distributed around the system. Artman and Garbis (1998) defined team SA as, 'the active construction of a model of a situation partly shared and partly distributed between two or more agents, from which one can anticipate important future states in the near future' (Artman and Garbis, 1998, p. 2).

Following on from Artman and Garbis (1998), Stanton et al. (2006) recently proposed the foundations for a theory of DSA. They contend that, during collaborative or distributed activity, cognitive processes (such as SA) occur at the system's level, rather than an individual level. Mirroring the approach taken by Artman and Garbis (1998), it is suggested that SA-related knowledge is distributed across the agents and artefacts (both human and non-human) comprising the system and that these knowledge 'themes' or 'topics', labelled information elements, represent what agents 'need to know' in order to achieve success during task performance. In this case, the term 'knowledge' represents the relationship between concepts (Shadbolt and Burton, 1995) and refers to task level knowledge, which relates to the goals and sub-goals of the task being performed. The ownership, usage and sharing of knowledge is dynamic and is dependent on the task and its associated goals. Agents therefore have different SA for the same situation, but their SA can be overlapping, compatible and complementary, and deficiencies in one agent's SA can be compensated by another agent. Stanton et al. define SA as 'activated knowledge for a specific task, at a specific time within a system' (Stanton et al., 2006), which echoes Bell and Lyon's (2000) presumption that SA can be defined as knowledge in working memory about elements in the environment. Stanton et al. propose then, that a situation requires the use of appropriate knowledge (held by individuals, captured by devices etc.) that relates to the state of the environment and those changes as the situation develops. The 'ownership' of this knowledge is initially at the system rather than individual level. This notion could be further extend to include 'meta-SA', where its knowledge of other agents' knowledge is contained in the system, such that each agent could potentially know where to go when they need to find something out. Stanton et al.'s model is described in full in Chapter 3.

Summary of Collaborative SA Models

The team SA models described above are evaluated in Table 2.2.

The review of team SA models presented in the literature leads us to conclude that there is a lack of a unified, universally accepted definition and theory of team SA. The approaches presented in the literature focus on either a summation of team member SA, a shared awareness of the situation, the overlap between team member SA requirements or on a distributed level of system awareness. It seems that currently there is a lack of a model that fully describes the processes (individual and team) involved, the content of team SA and also the factors affecting team SA.

Based on a synthesis of the literature, it is apparent that team SA comprises a team's collective awareness of the situation. Team members must possess SA related to their individual roles and goals within the team (some of which may be common to other team members), whilst also holding SA related to other team members, including an awareness of other team members' activities, roles and responsibilities, and also to the team overall, including goals and performance. SA-related data and knowledge are distributed around the team through team processes such as communication, coordination and collaboration, and serves to inform and modify team member SA, which is informed and modified by the overall teams SA. Thus, a tripartite composition of team SA is apparent: individual team member SA (some of which may be common or 'shared' with other team members), SA of other team members and SA of the overall

Table 2.2   Team SA theory comparison table

Theory	Domain of Origin	Domain Applications	Theoretical Underpinning	Process	Composition	Novelty
Team SA Model (e.g. Endsley & Robertson, 2000)	Aviation	Military, Aviation Maintenance	Three Level SA Model (Endsley, 1995a)	Perception of elements Comprehension of meaning Projection of future states Sharing of mental models	Individual SA Shared SA (Overlapping SA Requirements)	Team SA and Shared SA
Inter and Intra Team SA Model (Endsley & Jones, 2001)	Military	Military	Three Level SA Model (Endsley, 1995a)	Perception of elements Comprehension of meaning Projection of future states Sharing of mental models	Individual Sa Shared SA Inter Team SA Intra Team SA	Inter & Intra Team SA
Team SA Model (Salas et al, 1995)	Generic	None	Three Level SA Model (Endsley, 1995a) Teamwork Theory	Perception of elements Comprehension of meaning Projection of future states Team Processes	Individual SA Team Processes Information Seeking Information Processing Information Sharing	Team SA Processes
Team SA Model (Wellens, 1993)	Military	Military	Three Level SA Model (Endsley, 1995a) Distributed Decision Making Model (Wellens & Ergener, 1988)	Collection of raw data Application of decision rules Selection of plans Information space Situation Space Action Space	Information space Situation Space Action Space Communication Bridge	Distributed Decision Making Model
Distributed Cognition Approach (Artman & Garbis, 1998)	Teleoperations	Teleoperations	Distributed Cognition Theory (Hutchins, 1995)	Shared & Distributed Models	Partly Shared and Partly Distributed Model of Situation	Distributed Cognition Approach
Mutual Awareness Team SA Model (Shu & Furuta, 2005)	Process Control Artificial Intelligence	Process Control (DURESS)	Three Level SA Model (Endsley, 1995a) Shared Co-operative Activity Theory (Bratman, 1992)	Individual SA Mutual Awareness	Endsley's three levels Individual SA Mutual Awareness	Mutual Awareness Description of SA using heuristic rules
Distributed Situation Awareness Model (Stanton et al, 2006)	Maritime	Military, Maritime, Energy Distribution, Aviation, Air Traffic Control, Emergency Services, Driving	Distributed Cognition Theory (Hutchins, 1995) Distributed SA Theory (Artman & Garbis, 1998)	Individual SA Sharing of Knowledge Elements Team Processes	System Level Emergent Property Activated Knowledge Shared Knowledge	SA as an emergent property of collaborative systems

team. The three are extrinsically linked, of course, since individual team member SA includes the SA of other team members and of the team. It is therefore argued that, according to the most prominent literature, and at a simple level, team SA comprises three separate but related components: individual team member SA, SA of other team-members (task-work SA) and SA of the overall team (teamwork SA). Each of these forms of SA is affected by team processes and attributes. This is represented in Figure 2.6.

This combined awareness of one's own, other team members' and the overall team's situation is, according to the literature, what makes up 'team SA'. This perspective,

Table 2.2   *Concluded*

Measure	Process or Product?	Citation	Main Strengths	Main Weaknesses
SAGAT (Endsley, 1995b) SA Requirements Analysis (Endsley, 1993)	Product	7	1. Extension of the popular and widely applied three - level model - sound theoretical underpinning and lots of supporting literature 2. Widely applied in a variety of domains 3. Comes with prescribed SA measurement approach (SAGAT)	1. More of a simplistic extension of the individual three level model than a team model in its own right 2. Measurement is complex and impractical for real - world distributed tasks
SAGAT (Endsley, 1995b) SA Requirements Analysis (Endsley, 1993)	Product	5	1. Extension of the popular and widely applied three - level model - sound theoretical underpinning and lots of supporting literature 2. Considers Inter and Intra team SA 3. Comes with prescribed SA measurement approach (SAGAT)	1. More of a simplistic extension of the individual three level model than a team model in its own right 2. Measurement is complex and impractical for real - world distributed tasks
Individual SA Team Processes Compatibility of mental models TARGETS (Fowlkes et al, 1992)	Process & Product	10	1. Provides an insight into the team processes linked to team SA 2. Based on a review of teamwork literature 3. Relates model to team training and speculates on what to measure and how to measure it during team SA assessments	1. Measurement approach is more suited to assessing team behaviour and performance than SA and team SA measurement applications are scarce 2. The model is based on a review of the team literature rather than naturalistic or empirical study 3. Focussed more on team processes than on team SA
CITIES (Wellens, 1993) Post Task Questionnaire Task Performance	Process & Product	4	1. CITIES experimental paradigm developed specifically for assessing team SA 2. Discussion of effects of different communications media on team SA 3. Based on model of distributed decision making	1. SA assessments restricted to CITIES VR environment 2. Limited applications
Observation/Field Study	Process & Product	11	1. Systems level description that permits both individual, collaborative and systemic SA assessments 2. Sound theoretical underpinning	1. Limited applications 2. No prescribed measurement approach 3. Does not describe individual SA processes
TSA Simulation	Process & Product	1	1. Model attempts to describe the content of team SA and the behaviours involved in its development 2. Attempts to describe Team SA through the use of heuristic rules 3. Builds on existing SA theory and uses additional shared co-operative activity theory to present arguement	1. Complex description of team SA 2. Measurement approach is limited to authors domain 3. Limited application or validation
Propositional Networks (Stanton, Salmon, Walker, Baber & Jenkins, 2005)	Process & Product	1	1. Systems level description that permits both individual, collaborative and systemic SA assessments 2. Sound theoretical underpinning 3. Has been applied in a variety of collaborative domains	1. DSA description and measurement is subjective and often occurs post-task 2. Propositional Networks methodology lacks validation 3. Does not describe individual SA processes

along with other contemporary team SA theories, may be sufficient to describe team SA in simple, small-scale collaborative scenarios. However, for complex, real world collaborative scenarios, viewing and assessing team SA becomes somewhat more intricate. Take, for example, military Networked Enabled Capability (NEC) scenarios. Such tasks involve numerous agents and artefacts working both collaboratively and in isolation from one another whilst being dispersed geographically, often over great distances. Viewing and assessing team SA in such environments is acutely complex. The dispersed, real world nature of tasks in these environments inhibits (or at least makes impractical) the use of probe techniques such as the situation awareness global

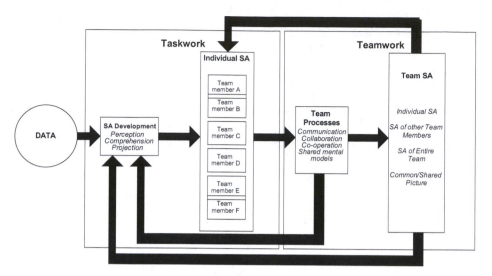

Figure 2.6   Team situation awareness

assessment tool (SAGAT) (Endsley, 1995b). The notion that team members share their awareness is also questionable since they are often distanced from one another and have different goals and roles within the team and also different levels of experience and schema.

A Note on Similarities with Other Concepts

Throughout the conduct of the literature review, many parallels were noted between SA and other concepts within the HF literature. For example, the concepts of situation assessment and sensemaking were all found to be similar and yet are distinct research areas in their own right. For the purposes of this literature review, it is worth touching on each concept and briefly discussing their parallels with SA.

Sensemaking is currently something of buzzword within HF circles and is receiving a great deal of attention (e.g. Endsley, 2004; Jensen, 2007). Brehmer (2007) even suggests that it has succeeded SA as everyone's favourite concept. Sensemaking refers to the process that people undertake in order to make decisions on how to act in situations that they encounter (Weick, 1995; cited in Jensen, 2007). According to Wieck (1993) 'the basic idea of sensemaking is that reality is an ongoing accomplishment that emerges from efforts to create order and make retrospective sense of what occurs' (Weick, 1993, p. 635). In a military context, Brehmer (2007) suggests that sensemaking is the process of understanding what needs to be done in order to accomplish a mission given the current situation. Alberts and Hayes (2007) suggest that 'sensemaking spans a set of activities that begins with developing SA and ends with preparing for action' (p. 34).

The concept has most notably been discussed at an organisational level (Endsley, 2004) although it has been described by some SA researchers as a different perspective on cognition (e.g. Dekker and Lutzhoft, 2004; cited in Endsley, 2004). It has also been discussed with regard to its role in the development and acquisition of SA. Endsley (2004), for example, suggests that sensemaking closely resembles a subset of the processes involved in SA, suggesting that it represents the process of developing level 2 SA (comprehension of meaning of elements) from level 1 SA (perception of elements in the environment) through 'effortful processes of gathering and synthesizing information, using story building and mental models to find some formulation, some representation that accounts for and explains the disparate data' (Endsley, 2004, p. 324). Endsley (2004) further suggests that sensemaking only represents a portion of the picture since it is essentially backward looking, whereas SA, as an ongoing model of the situation, is also focused on the future.

Alberts and Hayes (2007) suggest that 'sensemaking involves more than developing SA; it goes beyond what is happening to include what may happen and what can be done about it. This involves analysis and prediction, both of which require a model (mental or explicit) and the knowledge of or development of decision options that map to various alternative futures'.

It is clear that there are very close similarities between the two concepts. Sensemaking involves understanding what is required to achieve a goal of some sort given the current situation, whereas SA refers to the process of developing awareness of a situation and also the product of awareness that is developed. Endsley (2004) suggests that sensemaking research is complementary to SA research and is not directly at odds with it.

Some have also distinguished between the process of situation assessment and the resultant product of SA. Endsley (1995a), for example, defines situation assessment as 'the process of achieving, acquiring and maintaining SA' and suggests that it is necessary to distinguish between SA as a state of knowledge and situation assessment as the process used to achieve it. This is contradictory, since Endsley (1995a) refers to the processes of perception, comprehension and projection in her definition of SA. Viewing SA in this manner (as a product and nothing else) suggests that SA is a very static concept; rather, it is argued that SA should be viewed in terms of both the processes involved in its development and the end product in terms of what it comprises.

Conclusions

The importance of acquiring and maintaining appropriate levels of SA during task performance in complex sociotechnical systems has been proposed. Consequently, much research effort has been expended in order to identify how the acquisition and maintenance of SA works in complex systems, how to enhance the levels of SA acquired by individuals and teams during task performance and also to determine what factors have an impact on SA acquisition and maintenance. This literature review has attempted to provide a synthesis of the key ideas, concepts and theories related to SA that are presented in the literature.

Summary of Literature Review

The review indicated that, on the whole, the SA literature (both individual and team) is disparate. Many models of individual and team SA exist, but they present the construct quite differently from each another. The most prominent SA models are individualistic in that they focus on SA acquisition and maintenance solely from the point of view of individual operators in complex systems; they focus exclusively on the SA 'in-the-head' of individual operators and thus cannot satisfactorily be applied to describing SA during teamwork. Each of the individual models presented differs in terms of its treatment of SA as either a process or product or as a combination of the two and in its underlying psychological approach to SA. Despite the controversy, Endsley's three level model is by far the most popular and continues to drive research into the construct. None of the other individual models discussed has subsequently received significant attention in recent times. It is also notable that despite continued research, the construct is still overshadowed by the process versus product debate and some researchers even still question the extent to which SA represents a unique psychological construct in its own right (e.g. Bell and Lyon, 2000; Moray, 2004) rather than merely being a catch all term encompassing various elements of human cognition (e.g. perception, working memory and global workspace).

In light of the ever increasing use of teams in complex systems, many researchers have attempted to prescribe models of team SA (e.g. Endsley and Jones, 2001; Endsley and Robertson, 2000; Salas et al., 1995; Shu and Furuta, 2005; Wellens, 1993) and it is evident from the literature that there is still no universally accepted model. Ostensibly, team SA appears to be multi-dimensional and comprises individual team member SA, shared SA between team members and the combined SA of the whole team, the so-called 'common picture'. Most models focus on the summation of individual team member SA, the shared awareness of the situation, or the overlap between team member SA requirements. It seems, however, that there is a lack of a model that fully describes the processes involved, the content of team SA and also the factors affecting team SA. In particular, the interactions between team members and the impact on SA seem to largely have been ignored.

It is concluded, then, that SA from an individual operator perspective at least, albeit incongruently in most cases, is well described in the literature. In particular, Smith and Hancock's (1995) ecological account of SA as externally-directed consciousness comes closest to catering for the dynamic cyclical process of achieving and maintaining SA and presents the most logical and valid account of how SA 'works'. Most of the other models (e.g. Bedny and Meister, 1999; Endsley, 1995a) have some elements of truth in them. Team SA, on the other hand, remains a challenge, both in terms of its description and of its measurement and it is apparent that the area remains ill defined and requires much further investigation. This necessity for clarity in the area of team SA is enhanced by the increasing use of teams, technology and complex procedures in complex sociotechnical systems, which shows no signs of abating. Once the concept of team SA is clearly defined and described, only then can interventions be used to enhance team performance and safety in these systems.

Situation Awareness in Complex Collaborative Environments: The Way Forward

Existing individual, team and shared SA models, whilst each containing useful elements, may prove impractical when applied to the description and assessment of SA in complex, collaborative environments. Endsley's individual model has been extended to the team environment (e.g. Endsley and Robertson, 2000, Endsley and Jones, 2001); however, it has been argued that the three level, information processing-based model is somewhat inadequate when applied to collaborative systems. Aside from its individual operator orientation, Endsley's model is also beset by flaws which limit its utility within collaborative settings. Further, it will undoubtedly prove difficult to measure SA during real world collaborative activities using the SAGAT approach that is advocated by the three level model. Other models suffer due to their individualistic nature, their lack of a prescribed and practical measurement approach and their lack of empirical validation (e.g. Smith and Hancock, 1995; Bedny and Meister, 1999).

More recent literature describes the relatively new concept of DSA (e.g. Artman and Garbis, 1998; Stanton et al., 2006); these accounts build on the ideas surrounding distributed cognition expressed by Hutchins (1995) and differ from traditional cognitive viewpoints by analysing the interactions among people and among people and artefacts rather than the cognitive properties of individual people (Artman and Wærn, 1999). They are concerned with how information is represented and how representations are transformed and propagated throughout systems (Hutchins, 1995). It appears that DSA approaches are much more suited to describing and assessing the concept of SA within modern day collaborative systems, not least because they focus on the system itself as the unit of analysis and thus direct our attention beyond merely the cognitive properties of individuals onto the external representations and the interaction between them.

It is therefore concluded that, when examining SA in collaborative systems, systems oriented approaches, such as the DSA model proposed by Stanton et al. (2006) are the most suitable. These conclusions are based on a number of observations, not least that the models currently used are individually oriented and subsequently 'team SA' assessments often focus on SA as a cognitive construct and on its summation across team members. Systemic approaches, on the other hand, take the system itself as the unit of analysis and assess the construct as a systems endeavour. This permits the analysis of interactions and relationships at many different levels and allows a focus on specific interactions within subsystems. Viewing SA as a systems level emergent property is fruitful for a number of other reasons, including that it permits a systemic description of the knowledge comprising SA (which can be extrapolated to an individual SA level) and it allows judgements to be made on potential barriers to SA acquisition and maintenance. Further, considering SA in this way ensures that team SA within complex collaborative systems is viewed in its entirety, rather than as its component parts (i.e. individual team member SA). In such systems, tasks are rarely performed entirely independently of others, especially in complex situations and when critical decision-making is required (Artman and Garbis, 1998) – these activities tend to require coordinated activity between several individuals (Cannon-Bowers and Salas, 1990; cited in Salas et al., 1995). It is important therefore that team SA assessments consider this coordination in order to promote cohesion.

Due to their infancy, however, DSA models require further validation and extension through naturalistic study and experimental research. Additionally, exactly what these models mean in terms of interface and system design, team training, team performance enhancement and the design of collaborative systems and procedures needs further clarification. Thus, much further work is required in order to comprehensively describe the concept of SA in collaborative systems in terms of what it is, what it comprises, how it is acquired and what the factors affecting it are.

Chapter 3

How Do We Know What They Know? Situation Awareness Measurement Methods Review

Introduction

Putting theoretical concerns aside (see Salmon et al., 2008a, for a review), the level of SA that teams and operators possess is now recognised as a critical factor in system design and assessment (e.g. Endsley et al., 2003; Salmon et al. 2008a; Shu and Furuta, 2005). It follows then that researchers, practitioners and system, training and procedure designers need to be able to accurately describe and measure individual and team SA in these environments. The accurate measurement of SA is critical, not only to the advancement of SA-related theory, but also to artefact, system, procedure and training programme design and evaluation efforts. Researchers need valid and reliable methods of assessing SA in order to test and advance SA theory, whilst designers and practitioners need ways of assuring that SA is improved and not degraded by new artefacts, systems, interfaces, procedures or training programmes.

For this purpose a range of different SA measurement approaches has been developed by academics and practitioners. Analogous to the disparity between SA models, there is also great debate over what it is that these measures are actually measuring (i.e. SA or some other construct) and which of these measurement approaches are the most suitable for assessing SA in complex sociotechnical environments (e.g. Gorman et al., 2006). This debate is exacerbated further when considering the SA of teams working in collaborative environments (since the majority of approaches focus on individual operator SA measurement). The purpose of this chapter is to compare and contrast the different SA measurement approaches that are available to practitioners undertaking an assessment of the construct. Moreover, the review had the additional aim of identifying the most suitable approaches for describing and measuring SA in complex real world collaborative environments (with a view to selecting an appropriate measure for use during this research).

The methods review covered those SA measurement methods that are described in the academic literature and that are readily available for practitioners to use. The findings derived from the SA methods review are presented below.

Situation Awareness Measurement

As the construct of SA has become more and more eminent, attempts have been made to develop more sophisticated approaches for measuring SA. This has led to the development of a range of very different approaches. For example, in a review of SA measurement techniques, Endsley (1995b) describes a range of different approaches that had previously been used, including physiological measurement techniques (e.g. eye tracking devices, electroencephalograms etc.), performance measures (e.g. external task performance measures and imbedded task performance measures), subjective rating techniques (self and observer rating), questionnaires (post-trial and on-line) and freeze probe recall techniques (e.g. SAGAT).

A Note on the Reliability and Validity of Situation Awareness Measures

When discussing the selection and application of HF methods, their reliability and validity is a critical consideration. It is important to ensure that these techniques actually work. This seems an obvious statement to make and yet one of the consistent criticisms associated with our discipline is that some of the methods that HF practitioners use may not be reliable and valid (Annett, 2002; Stanton and Young, 2003). Stanton and Young (1999a, 2003) point out that despite the increased number of HF methods available, there is little evidence that the methods actually work. Further, in a review of over 90 HF design and evaluation methods, Stanton et al. (2005) only found validation evidence for a small number of the methods considered.

The objective way of testing whether or not HF methods actually work is to assess their reliability and validity (Stanton and Young, 1999a, 2003). In explaining the reliability and validity of HF methods, Stanton and Young (1999a) use the analogy of the accuracy of a rifle marksman. The *reliability* of his shooting refers to the grouping of the shots whereas the *validity* refers to the closeness of each shot to the centre of the target. The reliability of a method therefore relates to the repeatability of the results generated. Annett (2002), for example, suggests that reliability 'is about repeatability of results either by another observer or at different times under different conditions' (p. 229). A method is deemed reliable when it can be shown that it will generate the same results when used by different analysts, at different times and under different conditions. There are various forms of reliability that can be tested, including:

1. *Test-retest reliability.* Refers to the extent to which the method will produce the same results when used to measure the same participant over repeated tests under the same conditions.
2. *Inter-tester reliability.* Refers to the extent to which independent analysts will produce the same results when measuring the same phenomenon using the same method.
3. *Parallel forms reliability.* Refers to the extent to which two measures will produce the same results when used to assess the same phenomenon. It is essentially the degree of correlation between two independent measures of the same phenomenon.
4. *Internal consistency.* Refers to the consistency of the measure across items within a test.

The validity of a method, on the other hand, refers to the accuracy of the method in terms of what it is supposed to be measuring. Annett (2002) suggests, 'a method is regarded as valid if, after careful scrutiny, no objection or contradiction can be sustained' (p. 228). There are various forms of validity that can be tested, including:

- *Construct validity*. Refers to the extent to which the test defines the trait being measured (Annett, 2002).
- *Predictive validity*. Refers to the extent to which test scores is correlated with a score on a criterion test (Annett, 2002).
- *Face validity*. Refers to the extent to which a method appears to measure what it is supposed to be measuring as judged by an appropriate subject matter expert.
- *Concurrent validity*. Refers to the extent to which the results generated by a method correlate with the results produced by methods used to measure the same phenomenon.

Uhlarik and Comerford (2002) describes the following categories that should be considered when assessing the validity of SA measurement approaches:

1. *Face validity* – the degree to which the measure appears to measure SA as judged by a subject matter expert;
2. *Construct validity* – the degree to which the measure is underpinned by a sound theory or model of SA;
3. *Predictive validity* – the degree to which the measure can predict SA; and
4. *Concurrent validity* – the degree to which the measure correlates with other measures of SA.

For the purposes of this chapter, which focuses on team SA measurement during real world tasks, we are concerned with inter-tester reliability, the extent to which different analysts will generate the same data when measuring SA under the same conditions using the same method. That is, any method used should be reliable regardless of the analyst using it. In terms of validity, face validity, construct validity, predictive validity and concurrent validity are all applicable. Essentially validity in this case refers to the extent to which the measure is actually measuring SA and not some other psychological process or product.

Endsley (1995b) reports that when considering SA measurement techniques, it is necessary to establish that the technique:

1. measures SA and does not measure other processes or factors;
2. possesses the required level of sensitivity i.e. the technique can accurately detect changes in SA caused by novel technologies and programmes; and
3. does not alter SA during the measurement procedure.

The validation of HF techniques such as SA measurement techniques, whilst inherently necessary, is often neglected. This is for a number of reasons, mainly the high cost and resources invested when conducting validation studies. Stanton and

Young (1999a) also point out that researchers tend to stick with methods that they know and trust (often methods that they developed themselves) and so validation is assumed, rather than tested. In conclusion, to their exhaustive methods review, Stanton et al. (2005) reported that, of the methods available in the open literature, the majority are developed, subjected to an initial validation study and then discarded. Those techniques that are successful enough to be used elsewhere are often the techniques that receive the most validation testing. Typically, this leads to a small number of techniques within a particular area emerging as the most commonly used and extensively validated. For example, within the field of human error, the Systematic Human Error Reduction and Prediction Approach (SHERPA; Embrey, 1986) is by far the most commonly used human error identification (HEI) technique and has a large number of validation studies associated with it (Whalley and Kirwan, 1989; Kirwan, 1992; Baber and Stanton, 1996, 2002; Stanton and Stevenage, 1998). In the measurement of mental workload, the NASA-TLX (Hart and Staveland, 1988) is the most commonly used and the most widely validated of the various techniques available. The measurement of SA is no different, with the SAGAT (Endsley, 1995b) being by far the most commonly used approach and also the technique with the most associated validation evidence. For example, Jones and Kaber (2004) report that numerous studies have been conducted in order to assess the validity of SAGAT and that the subsequent evidence suggests that the method is a valid metric of SA. This highlights a potential problem: if new SA measurement techniques do not quickly catch on, there may not be any attempts to validate them. As a result, advances in the measurement of SA may become stilted, as practitioners will tend to use the most familiar methods available.

Situation Awareness Methods Review

The aim of the SA methods review was threefold: to identify and understand the range of available SA measurement methods available; to develop an in-depth guide for analysts and practitioners wishing to use the methods reviewed; and to evaluate the approaches in terms of their suitability for assessing SA during real world tasks in complex collaborative environments.

The measurement of team SA in complex sociotechnical systems poses a great challenge to the HF community. The environment is typically complex, dynamic and information rich and team members are often distributed across different geographical locations. Due to the collaborative and dispersed nature of team-based activity, an assessment of both individual and team SA is required in order to provide accurate measures of SA. As a result, any method that is used in such environments should possess the following three distinct capabilities:

1. *The ability to measure SA simultaneously at different geographical locations.*
   In order to gain a true measure of team SA, all the agents involved should be simultaneously assessed for their SA. However, due to the dispersed nature of collaborative activity, agents are typically remote from one another. Therefore, SA should be assessed at each of the different geographical locations involved. Any

technique used to assess team SA should therefore be capable of simultaneous administration at different locations. For example, the level of SA at different command locations (command centre, mobile units and foot units) may need to be assessed to ensure that the team involved has an adequate level of shared SA and that task relevant information is communicated efficiently. This would require a concurrent assessment of SA at the command centre, the mobile units and also commanders in the field.

2. *The ability to measure both individual and team SA.* Team endeavour comprises both teamwork and taskwork. Individual team members therefore have individual roles and possess individual goals, mental models and SA, whilst simultaneously pursuing team goals and maintaining a level of team SA. Any measure of team SA should be capable of describing and assessing SA both from viewpoint of the individual team members and the team as a whole.

3. *The ability to measure SA in real-time during real world tasks.* Typically, simulations of scenarios are used in order to assess SA. However, due to the dynamic, collaborative and dispersed nature of team-based activity, it appears that this may not be possible, and real world exercises conducted 'in-the-field' may be used. As a result, simulations of task scenarios and querying SA during task 'freezes' may not be appropriate. Team SA measures need to be applied during real world activities and so should be capable of assessing SA in real world collaborative environments.

An initial literature review was conducted in order to create a database of existing SA measurement techniques. The review identified references to over 30 different SA measurement techniques. A screening process was then employed in order to select the most appropriate techniques for further analysis. The screening process was based on technique availability, make-up and applicability to collaborative environments and was designed to quickly select or reject techniques from the initial database. As a result of the screening process, 18 SA measurement techniques were selected for further analysis (see Table 3.1).

In order to determine their suitability for use in the assessment of SA in collaborative environments each technique was evaluated using the following HF methods criteria (adapted from Stanton et al., 2005):

1. *Name and acronym* – the name of the technique and its associated acronym;
2. *Author(s), affiliations(s) and address(es)* – the names, affiliations and addresses of the authors are provided to assist with citation and requesting any further help in using the technique;
3. *Background and applications* – provides an introduction to the method, its origins and development, the domain of application of the method and also application areas in which it has been applied since its development;
4. *Team or individual* – denotes whether the measure was developed for the assessment of individual or team performance;
5. *Domain of application* – describes the domain that the technique was originally developed for and applied in;

Table 3.1   SA measurement techniques subjected to methods review

Method	Author/Source
SAGAT – Situation Awareness Global Assessment Technique	Endsley (1995a)
SA-SWORD – Subjective Workload Dominance metric	Vidulich & Hughes (1991)
SARS – Situation Awareness rating Scales	Waag & Houck (1994)
SART – Situation Awareness Rating Technique	Taylor (1990)
SALSA	Hauss & Eyferth (2002)
SABARS – Situation Awareness Behavioural Rating Scales	Endsley (2000)
PSAQ – Participant SA questionnaire	Endsley (2000)
SPAM – Situation-Present Assessment Method	Durso et al. (1998)
SACRI - Situation Awareness Control Room Inventory	Hogg et al. (1995)
C-SAS – Cranfield situation awareness scale	Dennehy (1997)
QUASA – Quantitative Assessment of Situation Awareness	Edgar & Edgar (2007)
CARS – Crew Awareness rating scale	McGuinness & Foy (2000)
MARS – Mission Awareness rating scale	Matthews & Beal (2002)
Verbal Protocol Analysis	Walker (2004)
Process Indices	Endsley (2000)
Performance measures	Endsley (2000)
CAST – Co-ordinated Awareness of Situations by Teams	Gorman, Cooke & Winner (2006)

6. *Procedure and advice* – describes the step-by-step procedure for applying the method as well as general points of advice;
7. *Flowchart* – presents a flowchart depicting the procedure that analysts should follow when applying the method;
8. *Advantages* – lists the main advantages associated with using the method;
9. *Disadvantages* - lists the main disadvantages associated with using the method;
10. *Example output* – presents an example, or examples, of the outputs derived from analyses with the method in question;
11. *Related methods* – any closely related methods are listed, including contributory and similar methods and/or methods to which outputs act as inputs (e.g. SA requirements analysis output is used to generate SAGAT queries);

12. *Approximate training and application times* – estimates of the training and application times are provided to give the reader an idea of the commitment required when using the technique;
13. *Reliability and validity* – any evidence on the reliability or validity of the method are cited;
14. *Tools needed* – describes any additional tools required when using the method;
15. *Bibliography* – a bibliography lists recommended further reading on the method and the surrounding topic area.

A number of the criteria used were intentionally descriptive, allowing the output to act as a user manual for each technique (e.g. background and applications, procedure and advice, flowchart etc). The full output from the methods review is presented in Stanton et al. (2005).

The methods assessed were divided into the following categories of SA measurement technique:

- SA requirements analysis;
- freeze probe techniques;
- real-time probe techniques;
- self-rating techniques;
- observer rating techniques;
- performance measures;
- process indices (e.g. eye tracker); and
- team SA measures.

A brief description of each of the different categories of SA measurement method and the methods reviewed within each category is presented below along with a discussion of the main advantages and disadvantages associated with each method.

SA Requirements Analysis

SA requirements analysis forms the first step in an SA assessment effort and is used to identify what exactly it is that comprises SA in the scenario and environment in question (i.e. before assessing operator SA one needs to understand what exactly it is that makes up that operator's SA in the situation under analysis). Endsley (2001) defines SA requirements as 'those dynamic information needs associated with the major goals or sub-goals of the operator in performing his or her job' (p. 8). According to Endsley (2001), they concern not only on the data that operators need, but also on how the data is integrated to address decisions. Matthews et al. (2004) suggest that a fundamental step in developing reliable and valid SA metrics is to identify the specific SA requirements of a given task. Further, Matthews et al. (2004) point out that knowing what the SA requirements are for a given domain provides engineers and technology developers with a basis on which to develop optimal system designs to maximise human performance rather than overloading workers and degrading their performance.

Endsley (1993) and Matthews et al. (2004) describe a generic procedure for conducting an SA requirements analysis that involves the use of unstructured interviews with subject matter experts (SMEs), goal-directed task analysis and questionnaires in order to determine relevant SA requirements. The output of SA requirements analysis then informs the development of the SA assessment technique used, since it specifies exactly what situational elements the operator should know about and understand in order to achieve SA during the task under analysis. Although other approaches such as Hierarchical Task Analysis (HTA; Stanton, 2006) have been used to conduct SA requirements analysis, Endsley's procedure is the only approach that has been developed specifically for this purpose.

Freeze Probe Techniques

Freeze probe techniques involve the administration of SA related queries on-line during 'freezes' in a simulation of the task under analysis. Typically, a task is randomly frozen and a set of SA queries regarding the current situation at the time of the freeze is administered. Participants are required to answer each query based on their knowledge and understanding of the situation at the point of the freeze. During the 'freezes' all operator displays and windows are typically blanked. For example, when assessing pilot SA, all cockpit displays (e.g. primary flight display, navigation display, altimeter, airspeed indicator etc.) and the aircraft windows are blanked. A computer is typically used to select and administer the queries and also to record the responses (although during low cost experimentation this is often done manually by researchers).

SAGAT (Endsley, 1995b) is the most popular freeze probe technique and was developed to assess pilot SA based on the three levels of SA postulated in Endsley's three level model. Although developed specifically for use in the military aviation domain, a number of different versions of SAGAT exist, including an air-to-air tactical aircraft version (Endsley, 1990), an advanced bomber aircraft version (Endsley, 1989) and an air traffic control version (Endsley and Kiris, 1995). Further, many freeze probe techniques based on the SAGAT approach have been developed for use in other domains. SALSA (Hauss and Eyferth, 2003), for example, was developed specifically for use in air traffic control. The SALSA queries are based on 15 aspects of aircraft flight, such as flight level, ground speed, heading, vertical tendency, conflict and type of conflict. The Situation Awareness Control Room Inventory (SACRI; Hogg et al., 1995) is an adaptation of the SAGAT and uses the freeze technique to administer control room based SA queries (it has also been used on-line as a real-time probe approach). SACRI was developed as the result of a study investigating the use of SAGAT in process control rooms (Hogg et al. 1995).

Freeze-probe techniques such as SAGAT (Endsley, 1995a) are the most commonly used SA measurement approaches. There are two primary advantages associated with freeze probe approaches. Firstly, they offer a direct measurement of operator SA, which removes the various problems associated with collecting post-trial and subjective SA data (see self-rating techniques summary). Participant responses provide information related to their understanding of the situation, which can then be compared to the actual situation at the point of the freeze. According to Endsley (2000), this provides a direct, objective measure of participant SA, since it directly assesses a participant's perceptions rather than inferring them from other behaviours. Secondly, the SAGAT approach (along with SART) is the most widely used and validated of the SA measures available and has consistently demonstrated reliability and validity in a number of domains (Jones and Kaber, 2004). According to Jones and Kaber (2004) numerous studies have been performed to assess the validity of the SAGAT and the evidence suggests that the method is a valid metric of SA. Further, Endsley (2000) reports that the SAGAT technique has been shown to have a high degree of validity and reliability for measuring SA. Collier and Folleso (1995) also reported good reliability for SAGAT when measuring nuclear power plant operator SA. In addition, in conclusion to a study of driver SA, Gugerty (1997) reported good reliability for the percentage of cars recalled; recall error and composite recall error. Regarding validity, Endsley et al. (2000) reported a good level of sensitivity for SAGAT, but not for real time probes and subjective SA measures. Endsley (1990) also reported that SAGAT showed a degree of predictive validity when measuring pilot SA, with SAGAT scores indicative of pilot performance in a combat simulation. The study found that pilots who were able to report on enemy aircraft via SAGAT were three times more likely to later kill that target in the simulation.

Whilst freeze probe techniques are the most popular and widely validated of existing approaches, they are also flawed in many ways, particularly when considering the measurement of team and distributed SA in complex real world environments. First, the use of freeze probe techniques 'in-the-field' is problematic and often proves impractical, if not impossible. Freezing a 'real' scenario (with multiple information sources) and administering SA queries to multiple agents who are dispersed across different geographical locations appears to be almost impossible. This limitation alone poses serious questions regarding the use of freeze probe techniques in real world collaborative SA assessments. Second, the intrusion into primary task performance caused by the task freezes is problematic. If a novel way of using freeze-probe techniques in the field were developed, then the intrusion into primary task performance would still presumably be high, which negates their use during real world tasks. Third, the focus of such approaches on participants' awareness of SA elements is problematic. Ostensibly, they make no allowance for experts' ability to achieve higher levels of SA without first achieving lower levels, or for the notion that experts may have parsimonious mental theories of the world (i.e. less could be more). This may mean that experts could be rated by such approaches as having poor SA even when they have a very efficient level of SA. Fourth, freeze probe approaches focus exclusively on SA in the heads of individual operators, which is not satisfactory when assessing SA at a team or systemic level. For example, individual operators who have no awareness of information of which other team members are aware would achieve poor SA scores when assessed using a freeze probe approach. However, it can be argued that the team or system has the level of SA that it requires for efficient task performance; team or system SA does not exist in the mind of individual operators, rather it resides in the interaction between them. Fifth, freeze probe techniques (and other approaches such as real time probe and subjective rating approaches) say very little about the processes used in developing and maintaining SA; whilst they allow analysts to assess what in the environment participants are aware of, they do not permit assessments of the processes involved in developing this awareness. Sixth, freeze probe techniques assume that a person who is 'aware' of more elements in the environment is a better one, however, in collaborative environments artefacts (e.g. interfaces, displays, whiteboards etc.) are typically used to 'remember' task related information for the team and so in this case SAGAT may rate individual team member SA as poor since they are not aware of everything. Finally, and perhaps most importantly, freeze probe approaches appear to ignore the mapping between SA elements; they assume that an operator who is aware of a set of pre-defined elements has good SA. This ignores the notion that SA could be more than the sum of its parts and that the linkage between SA elements (i.e. relationships between concepts) could determine the character of SA as much as the elements themselves.

There are, however, alternative approaches that could be used to remove the various problems of using a SAGAT style approach in the field. Instead of incorporating freezes, participants could be asked for their SA during low complexity portions of the task. Querying participants during natural breaks in the task (i.e. shift handovers) is another possible approach. In considering the measurement of SA in infantry operations, Endsley et al. (2000) report two alternatives designed to remove the problems associated with applying SAGAT in the field. The 'St Peter Technique' involves querying participants who have been 'killed' during task performance and the 'Angel of Death Technique' involves randomly selecting participants to be 'killed' and then immediately administering a series of SA queries. Both approaches, however, whilst allowing a freeze probe style approach to be applied in the field, are still problematic. The St Peter Technique may provide a biased measure of SA (Endsley et al., 2000) as those participants who are 'killed' during task performance may be those who possess lower levels of SA, and so participants with higher levels of SA may not be subject to measurement. Furthermore, both approaches still carry a high level of intrusion into the task under analysis.

Therefore, the use of freeze-probe techniques to measure team and DSA in collaborative systems is questionable. Even without the various flaws relating to how such approaches view SA (i.e. individual operator's awareness of elements), it is apparent that a novel variation of the freeze-probe technique designed to cater for the dispersed, collaborative nature of such environments requires development. Incorporating freezes into real world exercises represents a major challenge which has not yet been met by the techniques described in the open literature.

Real-time Probe Techniques

One alternative approach to the use of highly intrusive freeze probe techniques is the use of *real-time* probe techniques. Real-time probe techniques involve the administration of SA related queries on-line (during task performance), but with no freeze of the task under analysis. Typically, SMEs develop queries either before or during, task performance and administer them without a freeze at appropriate points in the task. Answer content and response time are typically taken as a measure of participant SA.

The Situation Present Assessment Method (SPAM; Durso et al., 1998) is a real-time probe technique that was developed for use in the assessment of air traffic controllers' SA. SPAM involves the use of on-line real time probes to evaluate their SA. The analyst probes participant SA using task related queries based on pertinent information in the environment (e.g. which of the two aircraft, A or B, has the higher altitude?) via telephone. The query response time (for those responses that are correct) is taken as an indicator of the operator's SA. Additionally, the time taken to answer the telephone is taken as an indication of participant workload (e.g. the longer the participant takes to answer the telephone, the higher their workload is assumed to be). SASHA (Jeannot et al. 2003) was developed by Eurocontrol™ for the assessment of air traffic controller SA in automated systems. SASHA comprises two techniques, SASHA_L (real-time probe technique) and SASHA_Q (post-trial questionnaire). SASHA_L is based on the SPAM technique (Durso et al., 1998) and involves probing the participant on-line using real-time SA related queries. The response content and response time is recorded. Once the trial is completed, the participant completes the SASHA_Q questionnaire, which consists of 10 questions designed to elicit subjective participant ratings of SA.

Based on a comparison of real-time probes, SAGAT and SART when used to measure operator SA in war and peace scenarios, Jones and Endsley (2000) reported that, when there is no simulation of the system under analysis and the task cannot be frozen, real time probes may provide a viable option for measuring SA.

The main advantages associated with real-time probe techniques are the reduced level of intrusiveness, since no freeze of the task is required (it is alleged that such techniques retain the direct, objective nature of freeze probe techniques whilst limiting the level of intrusion imposed on task performance), and the ability to be applied in-the-field during real world activities. However, the degree to which intrusion into task performance is reduced is certainly questionable. Whilst no freeze is required, the SA queries are still administered during task performance, which still represents a significant level of intrusion into the primary task. Furthermore, participant attention may be directed to the relevant SA-related information because of the query, which could bias the results obtained. Real-time probe techniques also suffer from a number of other major flaws. Due to the typically dynamic and unpredictable nature of collaborative tasks, the SA queries would presumably be generated in real-time, and not before task performance. The generation of probes in real-time would potentially place a great burden on SMEs used and may prove too difficult. Also, when using a real-time probe approach to assess team SA, numerous SMEs would be required due to the amount of personnel involved.

Real time probe approaches also suffer from the same criticisms as do freeze probe approaches regarding their focus purely on the SA levels held by individual operators and the way in which they view construct (i.e. failure to consider mapping between SA elements, failure to cater for SA as a feedforward phenomenon etc.). A valid measurement of team or shared SA would be difficult to obtain using such an approach.

Self-rating Techniques

Self-rating techniques are used to elicit subjective assessments of participant SA. Typically administered post-trial, self-rating techniques involve participants providing a subjective rating of the quality of their SA via a rating scale of some sort. The most commonly used self-rating approach, the Situation Awareness Rating Technique (SART;

Taylor, 1990), is a subjective rating technique originally conceived for the assessment of pilot SA. SART uses the following 10 dimensions to measure operator SA: familiarity of the situation, focusing of attention, information quantity, information quality, instability of the situation, concentration of attention, complexity of the situation, variability of the situation, arousal and spare mental capacity. SART is administered post-trial and involves the participant rating each dimension on a seven point rating scale (1 = low, 7 = high). The ratings are then combined in order to calculate a measure of participant SA. The 10 SART dimensions can also be condensed into the quicker three-dimensional (3-D) SART, which involves participants rating only attentional demand, attentional supply and understanding. The Situation Awareness Rating Scale technique (SARS; Waag and Houck, 1994) is a subjective rating technique that was developed for the military aviation domain. When using the SARS technique, participants subjectively rate their performance on a six-point rating scale (from *unacceptable* to *outstanding*) for 31 facets of fighter pilot SA. According to Waag and Houck (1994), the 31 behaviours represent those that are crucial to mission success. The SARS SA categories and associated behaviours were developed from interviews with experienced F-15 pilots (Waag and Houck, 1994). The 31 SARS behaviours are divided into eight categories representing phases of mission performance. The eight categories are: general traits (e.g. decisiveness, spatial ability), tactical game plan (e.g. developing and executing plan), communication (e.g. quality), information interpretation (e.g. threat prioritisation), tactical employment beyond visual range (e.g. targeting decisions), tactical employment visual (e.g. threat evaluation) and tactical employment general (e.g. lookout, defensive reaction). The Crew Awareness Rating Scale (CARS; McGuinness and Foy, 2000) technique has been used to assess command and control commander SA and workload (McGuinness and Ebbage, 2002). The CARS technique comprises two separate sets of questions based on Endsley's three level model (Endsley, 1995a). The content subscale consists of three statements designed to elicit ratings based on ease of identification, understanding and projection of SA elements (levels 1, 2 and 3 SA) during task performance. The fourth statement is designed to assess how well participants identify relevant task related goals in the situation. The workload subscale also consists of four statements which are designed to assess how difficult, in terms of mental effort, it was for participants to identify, understand and project the future states of the SA related elements in the situation. CARS is administered post-trial and involves participants rating each category on a scale of one (ideal) to four (worst) (McGuinness and Ebbage, 2002). The Mission Awareness Rating Scale (MARS; Matthews and Beal, 2002) technique is a development of the CARS technique (McGuiness and Foy, 2000) designed specifically for use in the assessment of SA in real world military exercises. The technique is normally administered post-trial, upon completion of the task or mission under analysis. The Quantitative Analysis of Situational Awareness (QUASA) technique (Edgar and Edgar, 2007) combines participant self-ratings with on-line probes in order to assess actual and perceived SA in military command and control scenarios. Participants are probed for their SA during task performance and simultaneously asked to rate their confidence in their answer to each SA probe. QUASA uses true or false probes and a confidence ratings scale (very low – very high) in order to assess actual and perceived participant SA. Finally, the Cranfield Situation Awareness Scale (C-SAS;

Dennehy, 1997) is a simplistic subjective rating scale that is used to assess student pilot SA during flight training exercises. C-SAS is administered either during or post-trial and involves participants rating five SA related components (pilot knowledge; understanding and anticipation of future events; management of stress, effort and commitment; capacity to perceive, assimilate and assess information; and overall SA). Each rating scale score is then summed in order to calculate an overall SA score.

The primary advantages of self-rating techniques are their ease of application (they are easy to use, quick to apply and incur a low cost) and their non-intrusive nature (since they are administered post-trial). However, subjective self-rating techniques are heavily criticised for a number of reasons, including the various problems associated with the collection of SA data post-trial (correlation of SA with performance, poor recall etc.), the extent to which people can be 'aware' of their own awareness (i.e. how do operators know that they were not aware of something?) and also issues regarding their sensitivity.

The use of self-rating techniques to measure SA during collaborative activity is attractive for a number of reasons. Firstly, self-rating techniques are non-intrusive on task performance, as they are completed post-trial. Secondly, they are very quick and easy to use and require very little training. Thirdly, because of their simplistic nature, very little cost is incurred when using self-rating techniques. Fourthly, and perhaps most importantly, self-ratings of SA can be obtained from different team members (Endsley et al., 2000) and so offer a potential avenue into the assessment of team SA, including the level of interaction between team members. Team members could potentially rate their own SA, the SA of other team members and the SA of the team as a whole. The majority of self-rating techniques are pen and paper tools, whereby participants rate their own SA on completion of the task under analysis, and so there is no requirement for expensive simulators, SMEs or a lengthy training process, all of which reduces the time and cost of the procedure considerably. The simplicity and low cost of self-rating techniques is reflected in their widespread use, with the SART technique (Taylor, 1990) being especially popular.

Despite the various advantages associated with the use of self-rating techniques, their potential use for measuring team and distributed SA is negated due to a number of flaws. Firstly, whilst the majority of self-rating techniques are generic and can be applied in numerous environments, a specific team SA approach is yet to emerge. Techniques such as SART, QUASA and MARS all focus on individual SA. Consequently, a team SA self-rating technique would require development, incorporating the dispersed collaborative nature of team-based activity. Secondly, there is a host of problems associated with the collection of SA data post-trial that would appear to rule out the use of a self-rating tool (on its own at least) for assessing SA in such environments. For example, previous research has indicated that SA ratings may be correlated with performance (Endsley, 1995b) i.e. a participant who performs well in a trial automatically rates their SA as good and, extended to the team level, a team who performs well would probably rate their SA as being of high quality. In addition, participants may be prone to 'forgetting' periods of the task when they possessed a poor level of SA and may more readily remember the periods when they possessed a superior level of SA. Endsley (1995b) reports that people are poor at reporting detailed information about past mental events and that post-trial questionnaires only capture participant SA at the end of the task under analysis. Thirdly, in various validation studies, the SAGAT (freeze probe) technique has proved to be superior in terms of reliability, validity and sensitivity when compared to the SART (self-rating) technique. Fourthly, as Endsley (1995b) points out, participant's ability to rate their own SA is questionable, as they may not be able to rate their poor SA accurately. Indeed it is questionable how accurately an individual can rate their own poor SA, as they may not realise that they have inadequate SA in the first place (i.e. can individuals really be aware of elements of the situation that they were not aware of?). Fifthly and finally, self-rating techniques typically do not reveal anything about the processes used to develop and maintain SA or about the content of SA during the task in question. Rather, self-rating approaches typically only offer a rating of how aware a participant felt they were during the task.

Observer Rating Techniques

Observer rating techniques are most commonly used to assess SA during real world tasks or tasks performed in the field. Observer rating techniques typically involve SMEs observing participants during task performance and then providing an assessment or rating of each participant's SA. The SA ratings are based on predefined observable SA related behaviours exhibited by participants during task performance.

The Situation Awareness Behaviourally Anchored Rating Scale (SABARS) is an observer rating technique that has been used to assess infantry personnel SA during field training exercises (Matthews et al., 2000; Matthews and Beal, 2002). The technique involves domain experts observing participants during task performance and rating them on 28 observable SA related behaviours. A five point rating scale (1 = very poor, 5 = very good) and an additional 'not applicable' category are used. The 28 rating items were designed specifically to assess platoon leader SA (Matthews et al., 2000).

Observer rating techniques are most commonly used when measuring SA in the field due to their non-intrusive nature and at first glance they appear to be suited to measuring SA during collaborative real world tasks: indeed, their non-intrusive nature and their ability to be applied during real world scenarios are the main advantages associated with the use of observer rating scales. However, on further investigation, it is quickly apparent that observer rating approaches are also beset by a number of flaws that may restrict their utility. The primary disadvantage associated with observer rating techniques concerns the construct validity of the measure. The extent to which observers can accurately rate participant SA is questionable (Endsley, 1995b), since the relationship between SA and task behaviours is ambiguous at best. Whilst there are observable behaviours that may indicate certain things regarding participant and team SA, the actual level of SA held cannot be accurately measured by observation alone. For example, a participant may exhibit appropriate behaviours even when they and the team have low SA, and a participant may not exhibit expected behaviours when they and the team have perfect SA. Similarly, the relationship between task performance and participant SA lacks clarity. Teams with poor SA can presumably achieve adequate task performance, whilst participants and teams with even a high level of SA could potentially perform poorly. In addition, it may prove extremely difficult to discriminate between different levels of SA across a team using observer ratings. Observer rating techniques may also be subject to bias, in that they may serve to alter participant behaviour. Knowing that they are being observed may change participant behaviour, in that they may strive to operate 'by-the-book' so to speak, and as a result the data obtained is subject to bias. Finally, observer rating techniques require repeated access to multiple SMEs over a long period of time, which is difficult to gain in most cases, especially in complex sociotechnical systems.

Performance Measures

Using performance measures to assess SA involves measuring relevant aspects of participant performance during the task under analysis. Depending on the task, certain aspects of performance are recorded in order to derive an indirect measure of SA. For example, in a military infantry exercise, performance measures may be 'kills', 'hits' or mission success or failure. When assessing driver SA, Gugerty (1997) used hazard detection, blocking car detection and crash avoidance as SA performance measures during a simulated driving task.

Performance measures are attractive as they are simple to obtain (since they are generated through the normal flow of the task and are normally recorded during experimentation or real world tasks anyway) and are non-intrusive to task performance. On the downside, using performance measures as a measure of SA is problematic for a number of reasons, but mainly due to the unclear relationship between SA and performance (see above). The main problem is the assumption that efficient performance is achieved because of efficient SA and vice versa. As referred to above, exemplary performance during a task does not necessarily point to a participant having good SA and vice versa. It may be that efficient performance is achieved despite inadequate levels of SA, or that poor performance is achieved regardless of a high level of SA. The unstable nature of the relationship between task performance and SA serves to diminish the suitability of performance measures as indicators of participant SA. The link between performance and SA clearly requires further investigation, and until this link is made clear, using performance as an indicator of participant SA is not acceptable.

Having said this, their use is not to be completely discounted as the procedure is often very simple, and the data may still have uses, particularly as a back-up SA measure to the other techniques employed.

Process Indices

Process indices can also be used to measure SA. Process indices refer to the measurement of the cognitive processes employed by participants in order to develop and maintain SA and involve recording these processes during task performance. Examples of process indices used to measure SA include eye movements (via eye tracking), verbalisations (via verbal protocol analysis) and communications (Endsley et al., 2000). The most commonly used process index is the measurement of participant eye movements and fixations using an eye-tracking device (e.g. Smolensky, 1993) in order to derive a measure of SA. Using this approach an eye-tracking device (e.g. FACELAB) is used to measure participant fixations during task performance, which can then be used to determine how the participant's attention was allocated to SA elements in the environment during the task under analysis. Another process index related methodology is concurrent Verbal Protocol analysis (VPA; Walker, 2004), which involves creating written transcripts of operator behaviour as they perform the task under analysis. VPA is used as a means of gaining an insight into the cognitive aspects of complex behaviours and is often used to indicate operator SA during task performance. The transcript is based on the operator 'thinking aloud' as they perform the task and the link to SA is based on the assertion that verbal expression of task-related information indicates that the participant has some awareness of it.

The main disadvantage associated with the use of process indices in general is their indirect nature; they tell us little about the product of SA held by the individual in question. Ultimately, the quality of SA is not assessed by such approaches. For example, one problem associated with the use of eye-tracking devices surrounds the 'look-but-failed-to-see' phenomenon (Brown, 2001). Whilst the eye-tracker data can point to which elements in the environment the participant fixated on, there is no assurance that the element in question was accurately perceived and so merely knowing that an individual looked at something does not confirm that they were aware of it. Furthermore, typical eye-tracking devices are temperamental in their operation and the data analysis procedure is a lengthy one, requiring great patience on behalf of the analyst.

Team Situation Awareness Measures

Interestingly, only relatively little attention has been given to the development of specific team SA measures, although this is on the increase at the time of writing. Initially, efforts to assess team SA involved scaling up individual SA measures to develop team SA variants (e.g. Bolstad and Endsley, 2007). The problem encountered is that each form of SA measure has its own significant flaws, most of which become exacerbated when they are applied to collaborative activities. Team SA measures tend to focus on the levels of overall team SA and/or the degree of shared awareness between members of a team and can be categorised into team probe-recall techniques, observer rating team SA techniques and team task performance-based SA assessment techniques. Team probe-recall techniques (e.g. Bolstad et al., 2005) involve the use of a SAGAT style approach in a team setting. This involves administering SA probes to all team members during freezes in task performance. These approaches suffer from the same criticisms that are aimed at SAGAT style approaches and are difficult to use during real world collaborative tasks (which are difficult to freeze and are often distributed over a wide geographical area). Typically, such approaches are used in a simulated environment. Observer rating team SA techniques involve SME observers observing team performance and rating each individual team member's level of SA and the level of team and shared awareness. Like their individual SA assessment counterparts, these approaches suffer from doubts over their validity, i.e. the extent to which observers can rate participant's internal levels of SA. The majority of team SA assessment techniques come under the umbrella of team task performance-based SA assessment techniques. Typically, responses to changes in the task and environment are used to assess how aware a team and its components are. The Coordinated Assessment of Situation Awareness of Teams (CAST; Gorman et al., 2006) is a recently developed approach that uses changes in the task environment to assess a team's SA. CAST uses situational 'roadblocks' and judgements on how the team responds to these roadblocks in terms of coordinated perception and action processes in order to derive a measurement of team SA. The main criticisms of this approach relate to the unclear relationship between performance and SA. Although a team may respond appropriately to a roadblock, their SA may have been diminished in some way. In addition, the CAST measure focuses exclusively on team SA and does not consider individual team member SA levels.

The main problem with most team SA measures is that they still focus on the measurement of individual team member SA; assessing each team member's SA and then making a judgement of the overall level of team SA is obviously problematic; as Salas et al. (1995) point out, there is much more to team SA than simply combining individual team member SA. These approaches do not take into account the interactions between team members or how SA is distributed across the human and technological system and thus are not truly measuring the collaborative concept of 'team' SA. For example, is an operator who is not aware of something of which he or she does not need to be aware, since another team member is aware of it or the information is held by a technological artefact (to which he or she can refer to when needed), really unaware? At the time of writing more system or team oriented approaches are emerging, such as the CAST approach, however as yet these have not yet received much attention within the literature.

A summary of the methods review is presented in Table 3.2.

In summary, the results of the methods review demonstrate that (aside from the SA requirements analysis procedure which would be required before any form of SA analysis), in their current format, existing SA measurement approaches are inadequate for the measurement of SA in complex collaborative environments. There are two main reasons underlying this conclusion. Firstly, the majority of the SA measurement techniques reviewed (all aside from the CAST approach) were developed specifically for the assessment of individual operator SA and thus do not cater for team SA. As Salas et al. (1995) pointed out, there is much more to team SA than merely combining individual team member SA and the literature review presented in Chapter 2 highlighted that team SA is a multidimensional construct that consists of individual team SA, compatible SA, shared SA, team processes and the interactions between team members. It is therefore clearly not acceptable to simply measure individual team member's SA and then aggregate it in order to derive an assessment of team SA; just because each team member has good SA does not mean that the team has good SA. The lack of specific team SA measurement approaches available in the literature was surprising; however, it is notable that the measurement of team SA is currently receiving increased attention from the HF community (e.g. Gorman et al., 2006; Stanton et al., 2006 etc.). Secondly, the review indicated that, when used in isolation, each of the different SA measurement approaches are beset by flaws that could potentially hinder any SA data obtained. For example, freeze-probe recall techniques are intrusive and cannot be applied 'in the field'

Table 3.2   Summary of SA measurement techniques review

Method	Method Type	Domain of Origin	Domains of Application	Individual and/or Team?	SMEs Required	Training Time
Crew Awareness Rating Scale (CARS; McGuinness & Foy, 2000)	Self rating technique	Military (infantry operations)	Military (infantry operations)	Individual	No	Low
Mission Awareness Rating Scale (MARS; Matthews & Beal, 2002)	Self rating technique	Military (infantry operations)	Military (infantry operations)	Individual	No	Low
Situation Awareness Behaviourally Anchored Rating Scale (SABARS; Matthews & Beal, 2002)	Observer rating technique	Military (infantry operations)	Military (infantry operations)	Individual	Yes	High
Situation Awareness Control Room Inventory (SACRI; Hogg et al, 1995)	Freeze probe recall technique	Nuclear Power	Nuclear Power	Individual	No	Low
Situation Awareness Global Assessment Technique (SAGAT; Endsley, 1995b)	Freeze probe recall technique	Aviation	Aviation Air Traffic Control Military Nuclear Power Driving	Individual	No	Low
SALSA (Huass & Eyferth, 2003)	Freeze probe recall technique	Air Traffic Control	Air Traffic Control	Individual	No	Low
SASHA (Jeannott, Kelly & Thompson, 2003)	Real time probe recall technique Post trial questionnaire	Air Traffic Control	Air Traffic Control	Individual	Yes	High
Situation Awareness Rating Scale (SARS; Waag & Houck, 1994)	Self rating technique	Aviation	Aviation	Individual	No	Low

whilst real-time probe techniques are difficult to apply and are still intrusive to primary task performance. Self-rating techniques suffer from a host of problems associated with collecting subjective SA data post-trial (e.g. correlation with performance, participant's inability to rate low periods of SA etc.) and the construct validity of observer rating techniques, process indices and performance measures is questionable.

Table 3.2 *Continued*

Application Time	Tools Required	Validation Studies	Main Strengths	Main Weaknesses
Low	Pen & Paper	Yes (2)	1) Developed for use in infantry environments 2) Less intrusive than on-line techniques 3) Quick, easy to use requiring little training	1) Construct validity questionable 2) Limited evidence of use and validation 3) Problems of gathering SA data post trial e.g. correlation with performance forgetting low SA periods
Low	Pen & Paper	Yes (2)	1) Developed for use in infantry environments 2) Less intrusive than on-line techniques . 3) Quick, easy to use requiring little training	1) Construct validity questionable 2) Limited evidence of use and validation 3) Problems of gathering SA data post trial e.g. correlation with performance forgetting low SA periods
Medium	Pen & Paper	Yes (2)	1) SABARS behaviours generated from infantry SA requirements exercise 2) Non-intrusive 3) Could potentially be adapted for use in team SA assessments	1) Extent to which observers can accurately rate internal construct of SA is questionable 2) The presence of observers may influence participant behaviour 3) Access to SME's and field settings is required
Medium	Task & System Simulation Computer	Yes (1)	1) Removes problems associated with collecting SA data post-trial 2) Direct approach 3) Gives an SA score for individuals based on their awareness of elements in the environment	1) Requires task and system simulation 2) Intrusive to primary task performance and may direct attention to SA elements 3) Cannot be applied during collaborative real world tasks
Medium	Task & System Simulation Computer	Yes (10+)	1) Direct approach 2) Extremely popular approach that has been subject to numerous validation studies 3) Removes problems associated with collecting SA data post-trial	1) Requires task and system simulation 2) Intrusive to primary task performance and may direct attention to SA elements 3) Cannot be applied during collaborative real world tasks
Medium	Task & System Simulation Computer	Yes (1)	1) Direct approach 2) Removes problems associated with collecting SA data post-trial 3) Based on the popular SAGAT approach	1) Requires task and system simulation 2) Intrusive to primary task performance and may direct attention to SA elements 3) Cannot be applied during collaborative real world tasks
Medium	Task & System Simulation Telephone Computer	Yes (1)	1) Offers two techniques for the assessment of SA 2) Administering probes in real -time removes the need for task freezes , and allows the technique to be applied during real world tasks	1) Probes may direct attention to required elements 2) Generation of appropriate SA queries places great burden upon analyst/SME. 3) Limited evidence of use or validation studies
Low	Pen & Paper	Yes (1)	1) Quick, low cost and easy to use, requiring little training 2) Non-intrusive to primary task	1) Problems of gathering SA data post trial e.g. correlation with performance forgetting low SA. 2) Limited use and validation evidence 3) Cannot be applied to team SA assessments

Conclusions

The purpose of this review was to identify and understand the different SA measures presented in the literature and to subsequently compare and contrast them in order to identify those approaches that are the most suitable for assessing SA during real world

Table 3.2   *Continued*

Method	Method Type	Domain of Origin	Domains of Application	Individual and/or Team?	SMEs Required	Training Time
Situation Awareness Rating Technique (SART; Taylor, 1990)	Self rating technique	Aviation	Aviation Air Traffic Control Military Nuclear Power	Individual	No	Low
SA-SWORD Vidulich & Hughes (1991)	Self rating technique	Aviation	Aviation	Individual	No	Low
Situation Present Assessment Method (SPAM; Durso et al, 1995)	Real time probe technique	Air Traffic Control	Air Traffic Control Aviation	Individual	Yes	High
SA Requirements Analysis (Endsley, 1993)	SA requirements analysis technique	Generic	Aviation Air Traffic Control Military Nuclear Power	Individual and Team	Yes	Med
Cranfield Situation Awareness Scale (C-SAS; Dennehy, 1997)	Self rating/ Observer rating technique	Aviation	Aviation	Individual	Yes	Low
Performance Measures (Various)	Performance measure	Generic	Various	Individual and Team	Yes	Low
Eye Tracker	Process indice	Generic	Various	Individual	No	Med
Quantitative Assessment of Situation Awareness (QUASA; McGuinness, 2004)	Probe/Self rating technique	Military	Military	Individual	No	Low
Verbal Protocol Analysis	Process indice	Generic	Military Driving	Individual and Team	No	Medium
Co-Ordinated Awareness of Teams (CAST; Gorman et al, 2006)	Team assessment method	Military	Military	Team	Yes	Low

collaborative activities. In conclusion, the review indicates that existing SA measurement techniques are inadequate for use in the assessment of team and distributed SA in complex sociotechnical systems. Whilst each form of technique (e.g. freeze-probe, real-time probe, self-rating etc.) possesses flaws which would hinder the data collected, the techniques also fail to meet the requirements specified earlier in the chapter, namely that any technique

Table 3.2 *Concluded*

Application Time	Tools Required	Validation Studies	Main Strengths	Main Weaknesses
Low	Pen & Paper	Yes (10+)	1) Quick, low cost and easy to use requiring little training 2) Generic - can be used in other domains 3) Non-intrusive to primary task performance and can be used during real world SA assessments	1) Problems of gathering SA data post trial e.g. correlation with performance forgetting low SA periods 2) Issues regarding sensitivity of the technique 3) Has not performed well in various validation studies and it is questionable whether it is in fact assessing SA or not
Low	Pen & Paper	Yes (2)	1) Quick, low cost and easy to use requiring little training 2) Useful when comparing two systems or artefacts 3) Generic and can be applied in any domain	1) Problems of gathering SA data post trial e.g. correlation with performance forgetting low SA periods 2) Does not provide a measure of SA 3) Limited application
Low	Task & System Simulation Telephone Computer	Yes (4)	1) No freeze required 2) Has shown promising results in validation studies 3) Administering probes in real-time removes the need for task freezes, allowing the technique to be applied during real world SA assessments	1) Low construct validity 2) Limited application 3) Attention may be directed to required SA elements
High	Pen & Paper Audio recording device	No	1) The output specifies the elements that comprise operator SA in the scenario under analysis 2) Output can be used to develop SA measure 3) The procedure is generic and can be applied in any domain	1) The procedure is time consuming involving observation, interviews and task analysis. 2) Access to numerous SME's is required for a lengthy period of time. This may prove difficult to gain 3) Describes only the SA elements and not the interactions between them
Low	Pen & Paper	No	1) Very quick, low cost and easy to use, requiring little training 2) C-SAS scales are generic, and can be applied in any domain 3) Can be used as a self-rating tool and an observer-rating tool	1) Unsophisticated measurement tool 2) No validation evidence associated with the technique. 3) Problems of gathering SA data post trial e.g. correlation with performance forgetting low SA periods
Low	Dependent upon task under analysis	No	1) Data collection is simplistic 2) Provides an objective measure if SA 3) Non-intrusive and can be applied during real world collaborative SA assessments	1) The relationship between performance and SA is an ambiguous one e.g. poor performance can still occur even when operators have poor levels of SA 2) Indirect assessment of SA 3) Suffers from diagnosticity and sensitivity problems
High	Eye Tracker equipment and software	No	1) Relatively unintrusive to primary task performance 2) Can be used to determine which environmental elements are attended to 3) Widely used	1) Equipment is temperamental and difficult to operate, cannot be used 'in-the-field' and the data analysis procedure is very time consuming 2) 'Look but do not see' phenomenon should be considered 3) Offers only an indirect assessment of SA (Endsley et al 2000).
Low	Pen & Paper	Yes (3)	1) Combines subjective ratings with SA probes. 2) Developed specifically for military command and control environments 3) Provides an assessment of actual participant SA and also their perceived SA (confidence in their SA)	1) Intrusive to primary task performance 2) Does not cater for teams. 3) Limited evidence of use and validation
High	Audio recording equipment	Yes	1) Verbalisations provide a genuine insight into cognitive processes 2) VPA provides a rich data source 3) Simplistic procedure that can be applied to teams during real world tasks	1) Data analysis procedure is extremely laborious and time consuming 2) Prone to bias. 3) Verbal commentary can sometimes serve to change the nature of the task
Low	Pen & Paper	Yes (1)	1) Developed specifically for team SA assessments 2) A novel approach that uses roadblocks to assess how teams react to situations – considers interactions between the team and the situation	1) It is questionable whether this approach can be applied during real world tasks 2) The relationship between responses to roadblocks and team SA is not clear 3) Access to SMEs required

used to assess SA during collaborative activity should be able to assess participant SA across multiple locations at the same time, assess both individual and team SA for the same task and also assess SA in real-time (i.e. during real world activities).

The methods review also produced a number of more general conclusions regarding the measurement of SA. Firstly, it was concluded that the SAGAT approach (Endsley,

1995a) is the most commonly applied SA assessment technique. Consequently, the SAGAT approach has the most validation evidence associated with it. Indeed, validation of the techniques remains problematic, and the literature review indicated that there has been only limited investigation into the validation of SA measurement methods. Aside from SAGAT and SART, both of which have been subjected to a significant number of validation studies, there is very limited validation evidence associated with existing SA measurement techniques.

Recommendations

The measurement of SA in complex collaborative environments poses a considerable but not unachievable challenge to the HF community. The concept of team SA requires much further investigation in itself, which in turn requires the provision of reliable and valid measurement procedures. This review indicates that, in their current format, existing SA measurement approaches are inadequate for this purpose and a novel approach is required. As highlighted previously, the main issues surrounding the measurement of SA in such environments are the need to assess both individual and team SA in real-time and at different locations simultaneously. From the categories of measurement technique available in the literature, not one can boast an ability to achieve this without incurring serious flaws that may hinder the data collected. There are two potential solutions to this problem. The first solution would be to develop a novel approach to the assessment of team SA that could satisfy these requirements. The second solution would be to combine the most successful SA measurement techniques in order to form a battery or 'toolkit' of SA measures. The lack of a single technique that can cope with both individual and team SA across multiple geographical locations in real time suggests that a toolkit of approaches may be suitable. A multiple measure approach ensures that SA data can be effectively crosschecked between measures in order to ensure reliability and accuracy. The concept of using a battery of human factors methods to achieve more efficient performance is not a new one. For example, in conclusion to a review of 38 existing human reliability analysis (HRA) and human error identification (HEI) techniques (Kirwan, 1998a), Kirwan (1998b) suggested that, as none of the techniques available satisfied all of the 14 criteria against which they were evaluated, a framework or toolkit approach using a mixture of independent HRA/HEI tools may be the most suitable approach to error analysis. It is also common to use a battery of methods (e.g. physiological measures, primary and secondary task performance measures and subjective measures) for the assessment of operator workload. A multiple measure approach has no doubt been used previously to measure SA, and it is not offered as a novel procedure; rather, it is offered as a solution to the considerable challenge faced when measuring team or distributed SA. The make-up of such an approach is unclear and considerable investigation is required in order to determine the logistics of such an approach.

Chapter 4

Distributed Situation Awareness: A New View on Situation Awareness in Collaborative Environments and its Measurement

Introduction

The inescapable conclusions from Chapters 2 and 3 are that currently there is a lack of an appropriate model of SA for complex collaborative environments and also that existing SA assessment methods are inadequate when considering the measurement of SA during real world collaborative activities. These findings combined confirm the assumption that our understanding of SA in such environments remains limited and subsequently serve to set the scene for the rest of this book; that is, it is our aim to further investigate the description and measurement of SA in collaborative environments. Our lack of knowledge regarding team SA acquisition and maintenance and the accompanying lack of approaches for measuring team SA is the first issue that should be addressed by this research. The purpose of this chapter is therefore to introduce new approaches to both problems. Firstly, a recently developed model of DSA, which accounts for SA in collaborative environments, is presented. Following this, the propositional networks methodology – a modelling approach that can be used to describe and assess DSA during real world collaborative tasks – is described. A simple command and control paradigm example is then used to demonstrate both approaches.

Distributed Situation Awareness

As pointed out in Chapter 2, the concept of DSA has recently emerged within the HF literature. DSA approaches are based on the notion that in order to understand behaviour in complex systems it is more useful take the system itself as the unit if analysis and to focus on the interactions between the parts of the system and the resultant emerging behaviour rather than study its parts in isolation (Ottino, 2003). DSA models are born out of distributed cognition theory (Hutchins, 1995) and Cognitive Systems Engineering (CSE; Hollnagel and Woods, 1999) approaches, both of which focus on the entire system rather than the individuals within in it as the unit of analysis when studying activity and cognition. These approaches take the view that the people and artefacts working within a system form a so-called 'joint cognitive system' and that cognitive processes emerge from and are distributed across this joint cognitive system. Cognition is therefore achieved through coordination between system units (Artman and Garbis, 1998).

DSA approaches argue that SA too exists at a systems level and can be viewed as an emergent property of collaborative systems. DSA is therefore taken to be the collective awareness of the entire system, a characteristic of the system in which the team is working (Artman and Garbis, 1998) and a product of the systems behaviour. This is, of course, in direct contradiction to those of models that treat SA as a uniquely internal cognitive construct (e.g. Endsley, 1995a); the approach instead views SA as a social phenomenon that exists in the artefacts and conversations around us. Artman (2000), for example, suggests that SA is 'not simply the sum of individual SA or a completely group level idea of a situation, it is an actively communicated and coordinated accomplishment between several members. This accomplishment emerges in a context where artefacts and information technology partly structure the possibility of sharing and distributing information' (Artman, 2000, p. 16).

Following on from Artman and Garbis's (1998) ideas (see Chapter 2), Stanton et al. (2006) recently proposed the foundations for a novel of theory of DSA, suggesting that DSA is a product of coordination between a system's elements and that the system collectively holds the SA required for task performance. Stanton et al.'s approach views knowledge as the relationship between concepts (Shadbolt and Burton, 1995) and suggests that SA-related information is held by and distributed between the agents and artefacts (both human and non-human) comprising the system. The combined sum of these information elements (or concepts) represents the system's DSA. Stanton et al. suggest that a system's awareness comprises a network of information elements and that the different agents comprising the system have different views of the network. These views are dynamic and change over time based on task requirements. DSA is defined as 'activated knowledge for a specific task, at a specific time within a system' (Stanton et al., 2006, p. 1291), meaning that the information required for SA becomes active (i.e. used) at different points in time based on the goals and activities being performed and their requirements. This definition has similarities with Bell and Lyon's (2000, p. 142) presumption that, 'SA could be defined as knowledge (in working memory) about elements of the environment' and also Cowan's (1988) description of activated working memory as awareness. Stanton et al. propose, then, that a situation requires the use of appropriate knowledge (held by individuals, captured by devices etc.) that relates to the state of the environment and those changes as the situation develops. The 'ownership' of this information is initially at the system rather than individual level. This notion is further extended to include 'meta-SA', where knowledge of other agents' knowledge is contained in the system, such that each agent could potentially know where to go when they need to find something out.

According to Stanton et al. (2006), each agent has unique but compatible (not shared) views on the situation. Each agent therefore plays a critical role in the development and maintenance of other agents' SA. Agents with limited or degraded SA can enhance or update their SA through interaction with another agent. This interaction between agents is critical to the maintenance of both the individual and DSA of the agents and the overall network involved. Stanton et al. (2006) point out that their approach does not contend that individual oriented perspectives are redundant; rather, they provide an alternative, but complementary approach to viewing and describing SA in collaborative systems. For example, in extending the DSA approach to Endsley's three level model, it is assumed that within collaborative systems, some individuals are engaged in perception tasks, some are engaged in comprehension and in the projection tasks and others are engaged in response execution tasks.

The main difference between individual and team models of SA and DSA approaches relates to the treatment of SA as a cognitive construct or as a systems construct. Most individual and team models suggest that SA exists in the mind of individuals whereas DSA approaches view SA as an emergent property or a product of the system itself. SA is therefore viewed as the 'glue' that holds the system together. Team and shared SA approaches also differ in that they view team SA either as a summation of individual SA or as the overlapping SA elements between team members. The key difference between existing team SA models (e.g. Endsley and Robertson, 2000; Salas et al., 1995) and the approach described by Stanton et al. relates to the issue of shared versus compatible SA and the treatment of SA as a systems level phenomenon. For example, Endsley (1989) and Endsley and Jones (2001) suggest that team SA comprises shared and team SA; shared SA refers to the level of overlap in common SA elements between team members and is defined as the degree to which team members have the same awareness of shared SA requirements (Endsley and Jones, 2001). Team SA, on the other hand, is defined as, the degree to which each team member holds the SA required for his or her function within the team (Endsley, 1989). Stanton et al.'s (2006) approach differs in that they view team SA as comprised of compatible SA rather than shared SA.

The notion of compatible SA requires further exploration. It may be that within collaborative systems, every team member does not need to know everything, rather, they possess the SA that they need for their specific task but are also cognisant of what other team members need to and do know. However, the extent to which systems should support 'sharing' of awareness is questionable. Although different team members may be aware of the same information, this awareness is not shared, since the team members often have different goals and tasks on which their view of the situation is based; they are often using information quite differently from one another. Each team member's SA is, however, compatible since it is different in content but is compatible in that it is all collectively needed for the overall team to perform the collaborative task successfully. Therefore, to suggest that all team members have their own SA and also shared SA with team members and the overall team could be an oversimplification. To use the analogy of a cog in a machine, every cog does not need to 'know about' all of the other cogs, rather, it only needs to be able to interact with those cogs adjacent to it – thus we propose that 'compatibility' is the key to team SA, rather than 'sharedness'. Any sharing of goals, intent and understanding arises out of the need of the individual team members to perform their tasks and not for its own sake. The ideas of 'sharing' have mutated into a vague belief that sharing ensures a cohesive team, but it can be argued that 'compatibility' leads to cohesiveness. DSA requirements are thus taken to be different from shared SA requirements (Stanton et al., 2006). Shared SA implies shared requirements and purposes whereas DSA implies different, but potentially compatible, requirements and purposes.

We contend that SA and team SA can be described more appropriately using the systems approaches advocated by Stanton et al. (2006) and Artman and Garbis (1998). It is also felt that the concept of 'compatible' SA is more appropriate than shared SA descriptions. Even in instances where team members have access to the same information it is apparent that factors such as the tasks being undertaken, roles within the team and past experiences ensure that their SA is significantly different. This, of course, has significant implications for collaborative system design, since it emphasises the need for disseminating only the appropriate information to the appropriate team members, rather than the need for team members to share their awareness.

Representing Distributed Situation Awareness

The departure from the study of information processing in the minds of individuals towards the study of systemic cognition brings with it the requirement for new approaches to model and assess SA. Individual-based measures such as SAGAT and SART are not applicable since they focus exclusively on the awareness 'in-the-head' of individual agents and overlook the interactions between them. Instead, what is required is an approach that is able to describe the concept from a systems perspective; this includes the information that is distributed around the system, the ownership and usage of this information by different system elements, how the information is combined together to form 'awareness' by different components of the system and the SA-related interactions between system elements. When this is requirement is coupled with the conclusions taken from the methods review presented in Chapter 3 (i.e. that the majority of SA measures are belied by flaws limiting their utility for assessing team SA) it is logical to either develop a new systems-based measure of SA or identify an existing distributed cognition-based approach that can be modified for DSA assessment.

Although the DSA approach takes much of its inspiration from distributed cognition theory (Hutchins, 1995) it is notable that the methods used in distributed cognition assessments may not be suited to SA assessments. Distributed cognition methods typically use ethnographic study (e.g. observational study and interview data) to develop basic textual descriptions of collaborative activity (e.g. Hutchins, 1995) and thus are not likely to provide the level of detail required for SA assessments. Clearly a more formal, systematic approach is required for DSA assessments.

We put forward the propositional network methodology (Salmon et al., 2008a; Stanton et al., 2009) as a way of describing a systems SA, since it depicts, in a network, the information underlying a system's knowledge, the relationships between the different pieces of information and also how each component of the system is using each piece of information. The approach has been applied to the analysis of collaborative work in a number of real world scenarios, including naval warfare (Stanton et al., 2006), railway maintenance operations (Walker et al., 2006), energy distribution substation maintenance scenarios (Salmon et al., 2008b) and military aviation airborne early warning systems (Stewart et al., 2008). A propositional network essentially comprises a network depicting the information underlying a system's awareness and the relationships between the different pieces of information. DSA is represented as information elements (or concepts) and the relationships between them, which relates to the assumption that knowledge comprises concepts and the relationships between them (Shadbolt and Burton, 1995). Depicting a system's awareness in this way permits representation of the usage of different pieces of information by different agents (human and non-human) within the system and also the contribution to the systems awareness by different agents. Propositional networks therefore provide a way of comprehensively describing the system's awareness and the information underlying it.

Representing Knowledge in Networks

The representation of knowledge in a network is not a new concept; semantic networks have been used by cognitive psychologists as a way of representing the association between items within a concept since the 1970s. Semantic networks are based on the long held belief that all knowledge is in the form of associations and represent concepts by depicting linked nodes in a network (Eysenck and Keane, 1990). Within a semantic network, each node represents an object, such as 'elephant' or 'mouse'. Each of the nodes within the network has associated properties, such as 'big' or 'small', 'tail' and 'trunk', 'mammal' or 'rodent'. The nodes are linked by pointers that are typically verbs such as 'has' or 'is'. The combination of the nodes, their properties and the links between them forms the semantic network. A simple semantic network is presented in Figure 4.1.

Similarly, concept maps (Crandall et al., 2006) have also been used to represent knowledge and do this via the use of networks depicting concepts and the relationships between them. According to Crandall et al. (2006), concept maps were first developed by Novak (1977; cited in Crandall et al., 2006) in order to understand and track changes in his students' knowledge of science. Concept maps are based on Ausubel's theory of learning (Ausubel, 1963; cited in Crandall et al., 2006) which suggests that meaningful learning occurs via the assimilation of new concepts and propositions into existing concepts and propositional frameworks in the mind of the learner. Crandall et al. (2006) point out that this occurs via subsumption (realising how new concepts relate to those

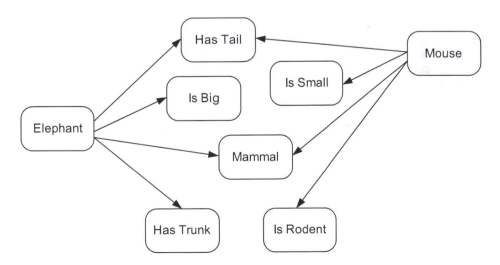

Figure 4.1   Example semantic network showing the associations between concepts

already known), differentiation (realising how new concepts draw distinctions between those already known) and reconciliation (of contradictions between new concepts and those already known). An example concept map of the concept map approach is presented in Figure 4.2 (adapted from Crandall et al., 2006).

With close similarities to both approaches, Anderson (1983) proposed the use of propositional networks to describe activation in memory. Propositional networks are similar in that they contain linked nodes; however, they differ from semantic networks in two ways (Stanton et al., 2006). Firstly, rather than being added to the network randomly, the words are instead added through the definition of propositions. A proposition in this sense represents a basic statement. Secondly, the links between the words are labelled in order to define the relationships between the propositions i.e. elephant 'has' tail, mouse 'is' rodent. Following Crandall et al. (2006), a simplistic propositional network about propositional networks is presented in Figure 4.3.

Constructing Propositional Networks

Propositional networks can be constructed from a variety of data sources, depending on whether DSA is being modelled (in terms of what it should or could comprise)

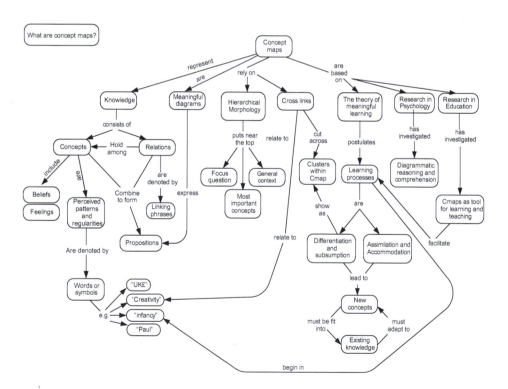

Figure 4.2   Concept map about concept maps

Source: Adapted from Crandall et al., 2006.

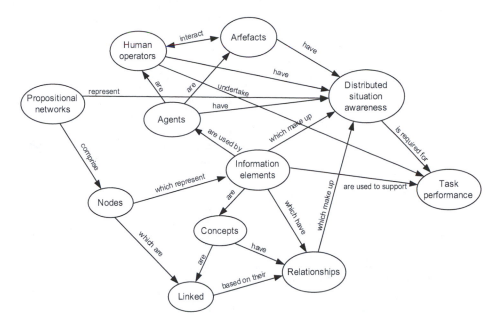

Figure 4.3   Propositional network diagram

or assessed (in terms of what it did comprise). These include observational or verbal transcript data, Critical Decision Method (CDM; Klein and Armstrong, 2004) data, HTA (Annett et al., 1971) data or data derived from work-related artefacts such as Standard Operating Instructions (SOIs), user manuals, procedures and training manuals.

In order to construct a propositional network, firstly the concepts need to be defined, followed by the relationships between them. For the purposes of DSA assessments, the term 'information elements' is used to refer to concepts. To identify the information elements related to the task under analysis, a simple content analysis is performed on the input data (e.g. verbal transcript, HTA or CDM responses) and keywords are extracted. These keywords represent the information elements, which are then linked based on their causal links during the activities in question (e.g. contact '*has*' heading, enemy '*knows*' plan etc.). The output of this process is a network of linked information elements; the network contains all of the information that is used by the different agents and artefacts during task performance and thus represents the system's awareness. These information elements represent what the system and the agents working within in it 'needed to know' in order to successfully undertake task performance. Information element usage can also be represented via shading of the different nodes within the network based on their usage by different agents during task performance. Thus, the information elements related to the DSA and also the usage, ownership and sharing of these information elements as the scenario unfolds over time can be defined.

A flowchart depicting the propositional network procedure is presented in Figure 4.4.

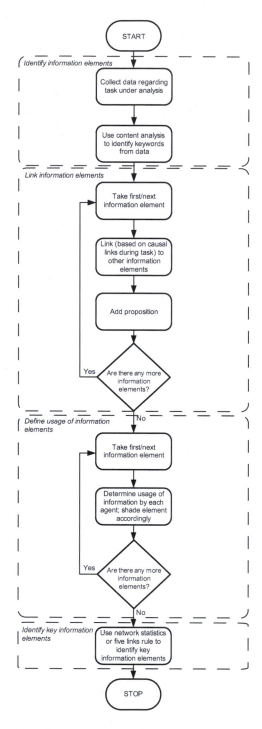

Figure 4.4   Propositional network procedure

To demonstrate how propositional networks are constructed, Figure 4.5 presents an extract of a verbal transcript collected during a study of DSA during land warfare activities (see Chapter 8); the keywords extracted from this transcript via content analysis are highlighted in bold and the ensuing propositional network is presented on the right hand side of the figure.

Distributed Situation Awareness Example

To familiarise the reader with both approaches, this section uses a simplistic command and control paradigm to demonstrate the DSA concept and the propositional network approach. The so-called 'sensor to effecter' paradigm has previously been used to model command and control scenarios (e.g. Jenkins et al., 2008) and was developed to represent a range of command and control domains (both military and non-military). Whilst it is accepted that this model is a simplified account of a sensor to effecter networks found in operational environments, the model does attempt to capture the essential features.

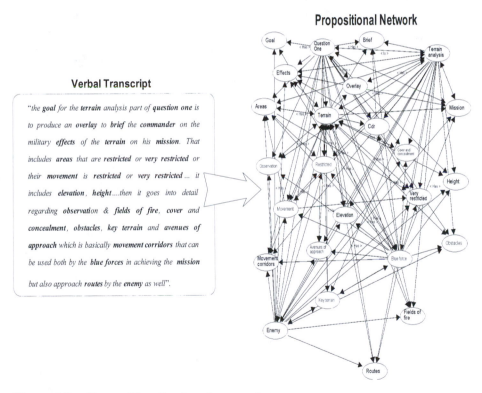

Figure 4.5 Propositional network example

Note: Figure shows verbal transcript and the resultant propositional network that is constructed based on the identification (via content analysis) of keywords from the verbal transcript data.

Other 'sensor to effecter' network analyses have used similar paradigms with some success (e.g. Dekker, 2003).

The paradigm environment is based in an urban setting of approximately 20 hectares. Within the environment, there are a number of concealed 'targets' that require the system's attention. There are two types of actors working within the environment. The first group of actors are reconnaissance units known as 'sensors'; sensors have the ability to sweep a geographic area and identify targets that need to be attended to. The second group of actors are 'effecters' who are responsible for attending to identified targets. In this simple paradigm, sensors are the only actors that can detect targets and effecters are the only actors who can attend to those targets identified.

There are a number of ways that information can be transmitted between the sensors and effecters and this is dependent upon the way that the system is configured. For the purposes of this example, a simple hierarchy containing sensors reporting to a commander and then the commander directing effecters is used. The command structure is presented in Figure 4.6.

The model structure presented in Figure 4.6 can be used to present a simple example of the DSA concept. In this case, there are four sensors located in the field, a commander situated at a remote command centre, and four effecters located in the field. The role of the 'system' is to locate and neutralise targets. The sensors role is to search, locate and identify targets, calculate their threat level and pass this information on to the commander. The commander's role is then to calculate the target priority (based on location, type, capability and threat), determine if the targets need to be neutralised and then allocate effecters to targets. The effecters' role is then to neutralise the targets and report their neutralisation back to the commander.

To begin with, the DSA of the system is incomplete. Although the system is aware of the possible presence of targets within the battlefield area, it does not know what or where the targets are or what their threat level and capability is. The sensors therefore search the battlefield area and on identifying a target record its location, classify the type

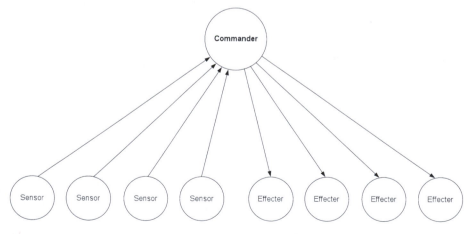

Figure 4.6   Example sensor to effecter command structure

of target and calculate its threat level. This information is then sent to the commander who takes the various target reports from each of the sensors and assigns a priority level to each target, determines whether or not they are to be neutralised and then identifies (based on effecter location and workload) an appropriate effecter to neutralise each target selected. Finally, the effecter, on being given the assigned target and relevant information (location, capability etc.) neutralises the target and confirms this with the commander.

This collaborative endeavour can be used to present examples of the DSA concept. The information required for the sensor to effecter system to work is distributed around its different components; each component holds information that is required for the system to work effectively. The sensor contributes information related to the targets identified (e.g. location, type, capability, threat etc.), the commander provides information related to the priority level of the targets, the targets that are to be neutralised and also the allocation of targets to effecters and, finally, the effecters provide information relating to the neutralisation of the targets. To demonstrate this, the SA requirements of the different components of the system are presented in Figure 4.7 overleaf. The SA is therefore distributed around the system; no one agent knows everything and for the system to work effectively every component's awareness needs to be combined with the others.

Each agent holds a different but compatible view of the situation based on their role and goals within the system. The sensors' SA comprises targets, their locations in the battlefield area and their capability and threat level. Each sensor's awareness is different to one another as they are located in different parts of the battlefield and are searching for different targets with differing levels of capability and threat. The commander's awareness comprises an overall picture of the targets, their location and threat levels, the priority and neutralisation requirement of the different targets and the availability and workload of the effecters; the commander uses the relationship between these factors to allocate targets to effecters. Finally, each effecter's awareness is based solely on the target(s) to which they have been assigned; they are aware only of the target in question, its type, capability and location and requirement for neutralisation. Thus, each component's view on the situation is entirely different but is compatible in that it is required collectively for the system to work.

The example presented can also be used to demonstrate the concept of 'SA transactions', which is proposed here as an alternative to the ideas of shared SA. For example, take a situation in which the commander has received information from the sensors regarding three targets, has assigned a priority to them and is currently looking at effecter locations and workload with a view to assigning targets to effecters. If at this stage the commander receives a new target report from a sensor in the field, the concept of SA transactions can be demonstrated. Here the commander receives a communication from the sensor in the field containing details regarding a fourth target, its type, location and capability and also a judgement on its threat level. This represents a 'transaction' or exchange in awareness between the sensor and commander; the sensor is passing on part of its awareness to the commander, who then combines the information received with his or her own awareness at that time. There is no sharing of awareness since both are using the information for their own ends; consequently

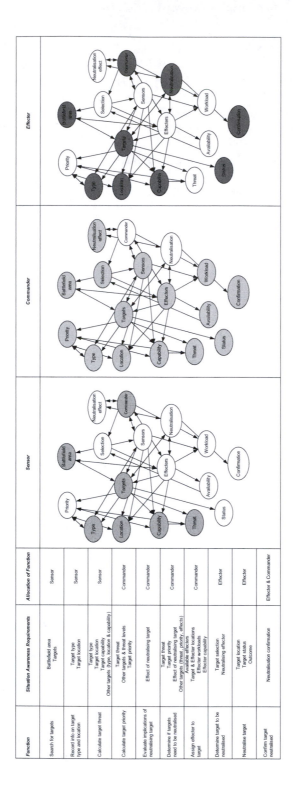

Function	Situation Awareness Requirements	Allocation of Function	Sensor	Commander	Effecter
Search for targets	Battlefield area Targets	Sensor			
Record info on target type and location	Target type Target location	Sensor			
Calculate target threat	Target type Target location Target capability Other targets (type, location & capability)	Sensor			
Calculate target priority	Target threat Other targets & threat levels Target priority	Commander			
Evaluate implications of neutralising target	Effect of neutralising target	Commander			
Determine if targets need to be neutralised	Target threat Target priority Effect of neutralising target Other targets (threat, priority, effects) Available effecters	Commander			
Assign effecter to target	Target & Effecter locations Effecter workloads Effecter capability	Commander			
Determine target to be neutralised	Target selection Neutralising effecter	Effecter			
Neutralise target	Target location Target status Outcome	Effecter			
Confirm target neutralised	Neutralisation confirmation	Effecter & Commander			

Figure 4.7   Sensor to effecter SA requirements example

their awareness, even when using the same information, is not the same. The sensor's awareness comprises simply the target, target type, location and threat level, whereas the commander's awareness of the new target also comprises its relationship with the other targets that he or she already knows about (i.e. proximity, threat level), priority level and potential effecter assignment. Thus, the sensor and commander in this example are not sharing their awareness; rather a transaction in awareness occurs between the two. The commander combines the new target information with his or her existing awareness, new relationships between concepts emerge and his awareness is modified as a result. This is represented in Figure 4.8.

In Figure 4.8, the commander's original picture includes an appreciation of the three targets present and an initial judgement on which effecters will deal with which of the three targets; on receiving a new report from one of the sensors containing information regarding another target (target, type, location, capability and threat) the commander's awareness is modified. The presence of new information is combined with his or her existing picture and target assignment is modified accordingly.

Comparison with Existing Models and Measurement Approaches

Distributed Situation Awareness Model versus Existing Models

The main criticism of existing SA models (see Chapter 2) in relation to this book is that they focus on the description of SA acquisition and maintenance from an individual

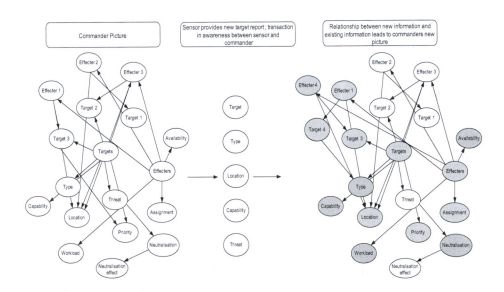

Figure 4.8   SA transaction and resultant modification of awareness

Note: The shaded information elements represent those elements that have been modified as a result of the transaction in awareness between the sensor and the commander.

perspective; further, the team models discussed either view team SA as a summation of individual team member SA or as a collection of individual and shared SA elements. It is our contention that these models maybe missing the point somewhat, that is, team SA is born out of the interactions between the agents comprising a collaborative system and thus is a social phenomenon that exists externally in the artifacts and interactions around us. Thus the main advantage of the DSA approach presented in this chapter over and above existing models is that it takes a systems perspective of the concept, which allows SA to be viewed in its entirety (rather than its component parts as other models permit) as a non-linear emergent property of collaborative systems. This in turn allows the collective information underlying DSA to be described, which in turn can be broken down further so that each agent's usage of, and contribution to, the information underlying DSA can be accounted for. The main benefit of this is that only coordinated activity can be considered and therefore true team SA is analysed. Many researchers have articulated the utility of studying collaborative activity from a systemic viewpoint (e.g. Hutchins, 1995; Hollnagel, 2001; Hollnagel and Woods, 1999; Ottino, 2003 etc.).

Propositional Networks versus Existing Situation Awareness Measures

It was concluded from the SA measurement technique review presented in Chapter 3 that existing SA measurement approaches are not suited to the analysis of SA during real world collaborative tasks. The basic requirement for such an approach was that it would be capable of assessing both individual and team SA simultaneously at different locations and during real world collaborative activities. The propositional network methodology satisfies each of these requirements. Firstly, propositional networks are capable of representing and assessing both individual and team SA since they can be used to describe the content and usage of an entire systems SA. This effectively allows analysts to understand what the content of the entire systems SA is and further what information (and how this information is related to other information) makes up the SA of each individual and each team operating within the overall system. Usage of the information elements (as identified through the content analysis on input data) is used to derive an assessment of each system component's SA. The mapping between information elements is also represented in the propositional networks, which allows the different views on the same situation to be described. Secondly, since the data used to construct propositional networks is collected via observational study, verbal transcripts or CDM interviews, propositional networks can be used to gather and assess SA data from different geographical locations, which in turn allows a simultaneous assessment of SA during the scenario under analysis to be made. Particularly useful here is the CDM approach, which is used to break the scenario in question down into key decision points, allowing SA data to be collected for each key decision point at each of the different locations involved. As outlined above, the propositional network has been used to assess SA during real world collaborative activities that involved teams of agents dispersed over geographical areas (e.g. Salmon et al., 2008; Walker et al., 2006). For example, the energy distribution scenarios studied in Chapter 5 involved teams comprising team members located in a central control room, a command control room, at various substations and also on the overhead lines. The data in this case was collected via observational study, verbal protocol analysis and CDM interviews at the different locations involved. Thirdly and finally, the approach can be used to assess SA during real world collaborative activities since the data collection procedure involved is not intrusive to task performance. The verbal transcript and HTA data is typically obtained via observational study and the CDM interviews are conducted post-task performance. In this way, it removes the need for scenario freezes and the administration of probes and does not require observer ratings of SA. It is acknowledged that in some cases the data may be subjective, since it can be provided by SMEs and analysts describing the task via HTA, but it is the our opinion that the combination of the two data sets (post-trial interview with and HTA developed by observers and validated by SMEs) negates the flaws typically associated with post-trial subjective SA assessments.

The propositional networks approach therefore differs from existing SA measurement approaches since it attempts to take a systems view of SA by linking and describing the SA-related information used by the system, sub-teams and agents involved in collaborative tasks. Whilst propositional networks do not attempt to quantitatively score each agent's SA quality (although this is something that could be introduced), they describe the content of the system's DSA during task performance and the usage of this information by the different agents involved. Judgements can be made on the quality of SA based on the information elements used.

Summary

The requirement for further clarification of the concept of SA in complex collaborative environments has been articulated by many researchers in the field (e.g. Artman, 2000; Gorman et al., 2006; Patrick et al., 2006; Salas et al., 1995; Salmon et al., 2006; Stanton et al., 2006; Shu and Furuta, 2005; Siemieniuch and Sinclair, 2006; Walker et al., 2006 etc.). We contend that the concept of DSA is suited to the description and evaluation of team SA in such environments. However, it is also recognised that the DSA concept is still very much in its infancy and subsequently that much further investigation and validation is required. In particular, more comprehensive descriptions of how DSA operates in collaborative environments, what processes are involved in its development and maintenance and how it can be augmented through system design and the use of training and procedures are required. Further, the sub-concepts of compatible and transactive SA introduced in this chapter require further exploration and explanation.

With regard to modelling SA, it is contended that the propositional network approach is more suitable than existing SA measurement approaches for describing and analysing SA during real world collaborative activities. However, it is also apparent that the approach requires further extension and validation as a measure of DSA.

In order to investigate the DSA concept and its measurement further and to formulate further explanation of DSA in complex collaborative environments, the following chapters of this book describe a number of case studies on DSA undertaken in both civilian and military complex collaborative environments. The primary aim of these studies was to further investigate the concept and to validate the theory and modelling approach described within this chapter.

Chapter 5

Distributed Situation Awareness in the Real World: A Case Study in the Energy Distribution Domain

Introduction

So far we have focused on SA theory and measurement and have identified a theoretical approach (DSA) and a modelling approach (propositional networks) to drive investigation into SA during real world collaborative tasks undertaken within complex sociotechnical systems. We now move from focussing on the current SA literature to the assessment of collaborative SA in such contexts. The aim from here on in is to investigate and extend the DSA theory and propositional network measurement approach outlined in the previous chapter, with a view to developing guidance on how to design systems, procedures and training programs so that DSA is enhanced and not inhibited.

The purpose of this chapter is to present the findings derived from a naturalistic study of DSA in the energy distribution domain, the aim of which was to investigate the DSA concept further and to attempt to provide empirical support for the DSA theory and to provide validation evidence for the propositional network methodology.

Energy Distribution Case Study

The propositional network methodology was used to analyse DSA in the UK energy distribution domain. This chapter focuses specifically upon two scenarios undertaken on a major UK electrical distribution network (a further two scenarios were also analysed). The distribution grid in question consists of 341 geographically dispersed substations in England and Wales, which are used to distribute electricity to consumers. Power stations (and feeds from continental Europe) energise the grid, which uses an interconnected network of 400,000 volt (400Kv), 275Kv (the super grid network) and 132Kv overhead lines and towers, or cables running in tunnels to carry electricity from source to substations. The substations are the national distribution company's interface with regional electricity companies who step down the grid's transmission voltages to 33Kv, 11Kv, 400v and 240v for domestic and industrial consumption. Although flexibly staffed, in operational terms they are remotely manipulated from a central control centre to ensure that the capacity available in the grid is used in optimal and rational ways, and that security of supply is maintained. Maintenance operations are also coordinated from another centre, thereby separating operations from safety. Two maintenance scenarios were analysed by the authors, using the propositional network approach. A brief description of each scenario analysed is presented below.

Scenario 1: Switching Operations Scenario

Scenario 1 took place at a substation in East London handling voltages and circuits from 275Kv down to 33Kv and a Central Operations Control Room (COCR). It involved the switching out of three circuits relating to so-called 'Super Grid Transformers' (SGT), which convert incoming transmission voltages of 275Kv down to 132Kv or 33Kv. Specifically, circuit SGT5 was being switched out for the installation of a brand new transformer for a bulk electricity consumer while SGT1A and 1B were being switched out for substation maintenance. Associated with such large pieces of high voltage apparatus are several control circuits, large overhead line isolators, remotely operated air blast circuit breakers, other points of isolation, compressed air equipment and oil cooling apparatus. All of this disparate equipment has to be handled and made safe in a highly prescribed manner by qualified personnel. In addition, the work had to be centrally pre-planned by electrical engineers to ensure that other circuits were not affected and that the balance and capacity of the system was not compromised. Qualified personnel work to these plans on site and liaise with the COCR at key points during this process.

Scenario 2: Maintenance Scenario

Scenario 2 took place at the COCR and a rural substation site. It involved the switching out of circuits and overhead lines in order to permit work to commence on pieces of the control equipment (current and voltage transformers), used to provide readings and inputs into other automatic, on site current, voltage and phase regulation devices. In addition, maintenance work was to be carried out on a line isolator (the large mechanical switching device that provides a point of isolation for a specific overhead line that departed from this substation (A) and terminated at another substation (B)) and major maintenance undertaken on a device called an earth switch. There were three main parties involved in the outage: a party working at Substation B on the outgoing substation A circuit, a party working at Substation A on the substation B circuit, and an overhead line party working inbetween the two sites.

In both scenarios, the COCR operator took on the role of network commander, distributing work instructions to the senior authorised persons (SAPs) and authorised persons (APs) located at the substations in the field. In addition to overseeing the activities that were analysed for the purposes of this research, the COCR operator was also involved in other activities being undertaken elsewhere on the grid and so had other responsibilities and tasks to attend to during the study. The other agents involved in the scenarios included the central command operator (CCO) and overhead line party (OLP) personnel working on the overhead lines. The COCR operator communicated with the other agents via landline and mobile phones. The COCR operator also had access to substation diagrams, work logs and databases and the internet. The network structure for scenario one is presented in Figure 5.1. The network structure for scenario two is presented in Figure 5.2.

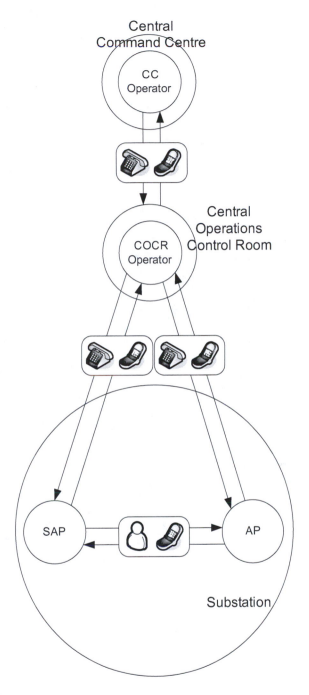

Figure 5.1   Scenario one network structure

Key: CC operator = central command operator; COCR operator = Central Operations Control Room operator; SAP = senior authorised person; AP = authorised person.

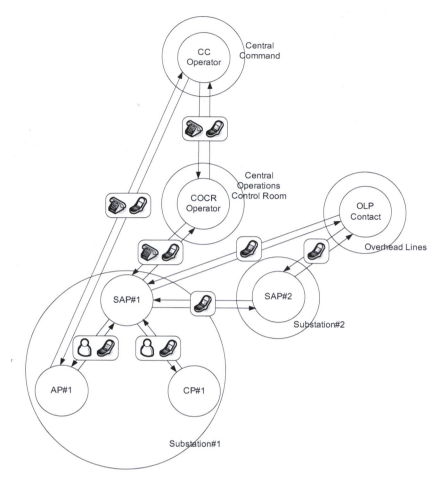

Figure 5.2   Scenario two network structure

Key: CC operator = central command operator; COCR operator = Central Operations
Control Room operator; SAP = senior authorised person; AP = authorised person;
CP = competent person; OLP Contact = overhead line party contact.

Methodology

Design

The study was an observational study that involved direct observation of the activities
undertaken during the scenarios analysed.

Participants

This study involved 11 participants who work for the energy distribution organisation in
question. Scenario one involved the following four participants: a CC operator, a COCR

operator and a SAP and AP. Scenario two involved the following seven participants: a CC operator, a COCR operator and a SAP, AP and CP at one substation, and SAP at another substation and an overhead line party contact. Due to access restrictions and the nature of the study (observation during real work activities), it was not possible to collect demographic data for the participants involved.

Materials

The observers used pen and paper, video and audio recording equipment to collect data during the observations.

Procedure

The analyses were based on data collected during live observational study of the two scenarios. In each scenario, two observers were located at the COCR observing the COCR operator and one observer was located in the field with the SAPs/APs at the substation where the work required was being undertaken. The analysts located at the COCR observed all the COCR operator's activities and were able to discuss the activities being undertaken and query different aspects of the scenarios as they unfolded. The analyst located at the substations observed the SAP/APs undertaking the work required and was also able to discuss aspects of the scenario with them. Observational transcripts were constructed and audio recordings were used to record the communications between those involved and any verbal protocol analysis data. The data recorded via observational transcripts included a description of the activity (component task steps, e.g. issue instructions to SAP at substation) performed by each of the agents involved, transcripts of the communications that occurred between agents during the scenarios, the technology used to mediate these communications, the artefacts used to aid task performance (e.g. tools, computers, instructions, substation diagrams etc.), time, and additional notes relating to the tasks being performed (e.g. why the task was being performed, what the outcomes were etc.). CDM interviews were conducted with the key agents involved (the COCR operator and the SAPs) upon completion of the scenario. This involved decomposing the scenario into a series of key decision points and administering CDM probes in order to interrogate the decision-making processes used at each point. The CDM probes used in this case are presented in Table 5.1. For validation purposes, a SME from the energy distribution company reviewed the data collected and the subsequent analysis outputs.

Results

Scenario One

Scenario one was divided into the following four phases: first issue of instructions; deal with switching requests; perform isolation; and report back to COCR. Propositional networks were constructed for scenario one using the observational transcripts, CDM interview responses and HTA of the tasks performed. The overall propositional network

Table 5.1   CDM probes

Goal Specification	What were your specific goals at the various decision points?
Cue Identification	What features were you looking for when you formulated your decision? How did you know that you needed to make the decision? How did you know when to make the decision?
Expectancy	Were you expecting to make this sort of decision during the course of the event? Describe how this affected your decision making process.
Conceptual	Are there any situations in which your decision would have turned out differently?
Influence of uncertainty	At any stage, were you uncertain about either the reliability or the relevance of the information that you had available?
Information integration	What was the most important piece of information that you used to formulate the decision?
Situation Awareness	What information did you have available to you at the time of the decision?
Situation Assessment	Did you use all of the information available to you when formulating the decision? Was there any additional information that you might have used to assist in the formulation of the decision?
Options	Were there any other alternatives available to you other than the decision you made?
Decision blocking – stress	Was their any stage during the decision making process in which you found it difficult to process and integrate the information available?
Basis of choice	Do you think that you could develop a rule, based on your experience, which could assist another person to make the same decision successfully?
Analogy/ generalisation	Were you at any time, reminded of previous experiences in which a similar/different decision was made?

Source: O'Hare et al., 2000.

for scenario one is presented in Figure 5.3 and the propositional network depicting information element usage is by the agents involved is presented in Figure 5.4. Figure 5.3 provides an overall picture of the information elements used by the 'system' during scenario one; it therefore provides an overview of the system's awareness during scenario one. Figure 5.4 presents an example representation of the usage of the different information elements by each of the agents involved (represented via shading of the information elements). The enlarged portion of the network depicts an example of

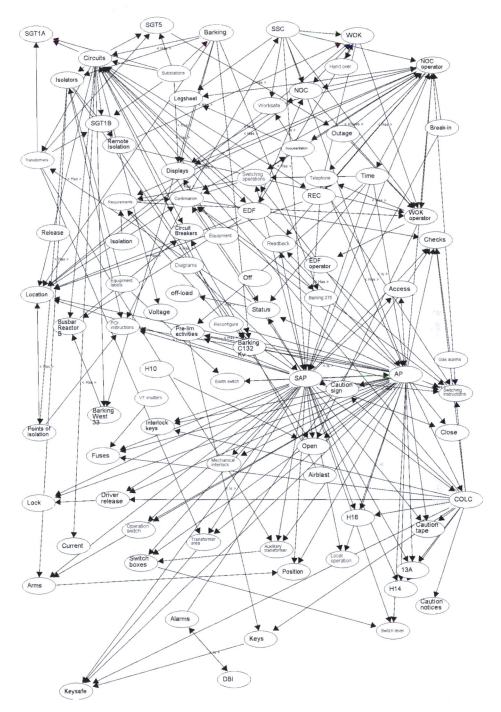

Figure 5.3   Propositional network for scenario one

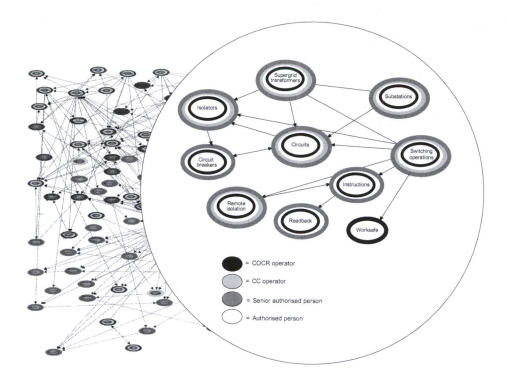

Figure 5.4   Propositional network for scenario 1 with blown up extract showing example information element usage by each agent involved

the information element usage by the different agents involved. The total number of information elements used by each agent during scenario one was calculated using the propositional networks. The frequency of information element usage by the different agents during scenario one is presented in Figure 5.5.

The graph presented in Figure 5.5 shows that the AP and SAP used the highest number of information elements (77) over the course of scenario one, followed by the COCR operator who used 53 and the CC operator who used 44. Figure 5.5 suggests then that the SAP and AP needed more information elements for their specific tasks.

Scenario Two

Scenario two was divided into the following phases: prepare plant, personnel and apparatus for proposed activity; issue instructions; perform isolation tasks; apply earthing; report operations complete; issue permit, and demark isolated equipment. Propositional networks were constructed for scenario two using the observational transcripts, CDM interview responses and HTA of the tasks performed. Propositional networks were constructed and the usage of the different information elements by different agents throughout the scenario was calculated.

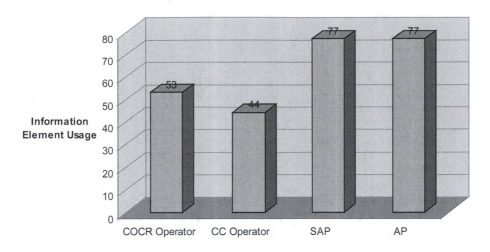

Figure 5.5   Information element usage during scenario one (overall)

The propositional networks were used to calculate the total number of information elements used by each agent, both overall during scenario one and during each scenario phase. The total number of information elements used by the different agents during scenario two is presented in Figure 5.6. The total number of information elements used during each of the seven phases comprising scenario two is presented in Table 5.2.

The graph presented in Figure 5.7 shows that the AP and SAP used the highest number of information elements (91 and 84 respectively) over the course of scenario two, followed by the COCR operator who used a total of 67 information elements, the CC operator (57) and finally the OLP contact, who used the least amount of information

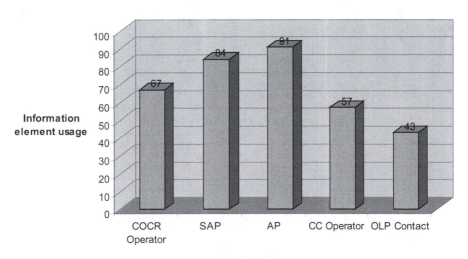

Figure 5.6   Information element usage during scenario two (overall)

Table 5.2   **Frequency of information element usage per phase during scenario two**

Agent	Information element usage							
	Scenario phase							
	1	2	3	4	5	6	7	Total
COCR Operator	22	41	24	0	20	30	0	**67**
SAP	34	36	0	29	18	37	0	**84**
AP	33	29	0	24	18	3	14	**91**
CC Operator	26	0	0	0	0	0	0	**57**
OLP Contact	15	0	0	9	13	0	0	**43**

elements (43). Again, Figure 5.7 indicates that the SAP and AP needed more information to support their activities during scenario two.

Table 5.2 presents a breakdown of information element usage during each of the seven phases identified. Table 5.2 shows that the COCR operator used the most information elements during phases 2 (issue of instructions), 3 (perform isolation tasks) and 5 (issue of permit). The SAP and AP used the most information elements during phases 1 (prepare plant and personnel for activity), 4 (apply earthing), 6 (demarking of equipment; SAP only) and 7 (prepare for maintenance activities; AP only). Table 5.2 provides an indication of how the usage of information varies over the course of a collaborative task and also the dynamic nature of DSA.

The key information elements were also identified for each scenario using sociometric status and centrality statistical calculations. Sociometric status provides a measure of how 'busy' a node is relative to the total number of nodes present within the network under analysis (Houghton et al., 2006). In this case, sociometric status gives an indication of the relative prominence of information elements based on their links to other information elements in the network. Centrality is also a metric of the standing of a node within a network (Houghton et al., 2006), but here this standing is in terms of its 'distance' from all other nodes in the network. A central node is one that is close to all other nodes in the network and a message conveyed from that node to an arbitrarily selected other node in the network would, on average, arrive via the least number of relaying hops (Houghton et al., 2006). Key information elements are defined as those that have salience for each scenario phase, salience being defined as those information elements that act as hubs to other knowledge elements. Those information elements with a sociometric status value above the mean sociometric status value and a centrality score above the mean centrality value were identified as key information elements. The key information elements for the two scenarios analysed are presented in Figure 5.7.

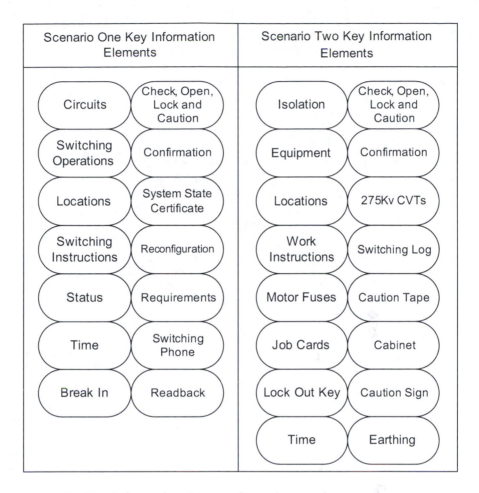

Scenario One Key Information Elements		Scenario Two Key Information Elements	
Circuits	Check, Open, Lock and Caution	Isolation	Check, Open, Lock and Caution
Switching Operations	Confirmation	Equipment	Confirmation
Locations	System State Certificate	Locations	275Kv CVTs
Switching Instructions	Reconfiguration	Work Instructions	Switching Log
Status	Requirements	Motor Fuses	Caution Tape
Time	Switching Phone	Job Cards	Cabinet
Break In	Readback	Lock Out Key	Caution Sign
		Time	Earthing

Figure 5.7   Key information elements for each scenario

Discussion

This chapter has presented the results derived from an analysis of two energy distribution maintenance scenarios. The purpose of this study was to build on earlier descriptions of DSA theory and DSA measurement and investigate the notion that SA in collaborative systems can be effectively described from a systems perspective, rather than as a summation or sharing of individual team member SA. Further, it was our intention to further investigate the concept of DSA within a complex, real world setting.

The findings are discussed firstly in relation to the structure, quality and content of the energy distribution network's DSA and also the DSA theory outlined in Chapter 4. In the scenarios analysed, the DSA of the networks involved was adequate to support efficient, timely and safe task performance, since all operations were completed successfully as envisaged and without incident. The COCR, who took the role of network commander, was provided with adequate SA by the system in order to plan, monitor, control and coordinate operations in the field, whilst the SAPs and APs who undertook the majority of the work also held adequate levels of SA to undertake their required tasks efficiently. Based on our analysis, it is apparent that the high quality of the networks DSA in the scenario's observed was a function of four factors: the efficient *communications* links between the agents involved, the use of well thought out and rigidly adhered to *procedures*, the *structure* of the network itself and also the clarity of *role* definitions.

The communications links available allowed DSA to propagate efficiently through the network of agents involved. For example, each of the agents involved had one or more ways of communicating with the other agents in the network, such as landline telephone, mobile phone and emails. Obviously, when collaborating across distances, SA is mediated by technology (Sonnenwald et al., 2004) and previous research on team SA has highlighted the importance of efficient communications links in collaborative systems (e.g. Stanton et al., 2006; Gorman et al., 2006). However, the procedures used also encouraged continual, explicit communication between each of the agents involved. For example, the procedures dictated that, on writing the work instructions, the COCR operator would contact the SAP or AP in the field to issue the work instructions. This involved the COCR operator reading the instructions one-by-one and the SAP or AP making a note of them and then reading them back to the COCR operator to confirm receipt. The procedures also made work progress updates compulsory (i.e. provided to the COCR operator by the SAP and AP in the field). The structure of the network (along with the procedures adopted) also facilitated DSA since the hierarchical organisation meant that the COCR operator, effectively as network commander, would contact (or be contacted by) agents in the field in order to gather work progress updates, a process which served to update the DSA of the system throughout the activities. Finally, since the roles within the networks were so clearly defined, the 'meta SA' of the system (knowledge of other agents' knowledge) was facilitated, and so when SA-related knowledge was required the agents involved knew where to go to get the required information.

In both scenarios, however, certain characteristics of the network's DSA were noteworthy. At times, the DSA appeared to be 'out-of-date' or at least lagging behind the real state of the world. Due to the dispersed nature of the networks and the fact that the agents in the field could often become un-contactable (either working away from telephones or out of range of mobile phone reception), the COCR would not have an up-to-the-minute level of SA and thus his SA would be based on the last situational report that he had received. This meant that the joint picture could often be dated. This, however, did not prove detrimental to the network's performance of the tasks required and could often be corrected through additional communications between the agents in question. Also, the unstable nature of some of the communications links, combined with a lack of specific agent geographical location information, could often mean that, in the event of not being able to contact the agents in the field, the COCR operator did not know exactly where the agents were located and thus exactly what they were doing.

In both scenarios, DSA and the individual SA of each agent was updated and maintained via explicit and implicit interaction between agents. Within the networks,

DSA was 'facilitated' primarily by the COCR operator, who continuously developed and maintained the overall 'big picture' via information exchanges with other agents (human and technological) in the network. This allowed him to collect and integrate information from a number of different sources, including operators in the field (SAPs, APs and overhead line parties), substation diagrams, worksheets, computers and databases. The COCR operator then distributed the 'joint picture' throughout the network via verbal communications. In this way, the COCR operator effectively acted as the network hub in terms of DSA maintenance.

Although the COCR was the primary DSA facilitator, the critical role of each agent in updating and maintaining other agents' SA and thus the DSA of the entire network was also demonstrated. For example, there were times during each scenario when the COCR operator's SA was incomplete and was only updated by communications, both incoming and outgoing, with the other agents in the network, and by interactions with other data sources at the COCR, such as displays, worksheets, computers and databases. These findings further highlight the importance of communication links in the maintenance of DSA.

Perhaps of most interest to DSA theory was the fact that *compatible* SA, rather than *shared* SA, was extant during the activities observed. It was found that each agent's SA was not the same but was in fact different and thus not shared, which goes against the existing 'shared SA' view of how SA works in collaborative environments (e.g. Endsley, 1995a; Endsley and Robertson, 2000; Endsley and Jones, 1997). In this case, it was apparent that each agent involved had different but requisite SA during the activities, and thus it was concluded that the different agents involved held different but *compatible*, rather than shared, SA for the same situation. For example, the COCR operator held a very high level overall picture, including the current status of the overall work scenario and general knowledge relating to the various work being conducted by each of the parties involved (e.g. who was doing what, why they were doing it and what they would be doing next), whilst the agents in the field (e.g. SAPs and APs) mainly held specific SA related primarily to the work that they were currently engaged in and their individual goals. The COCR knew what activities the SAPs and APs in the field were undertaking and thus what they were aware of, but he did not have a detailed and dynamic awareness of their activities. Thus, although each agent held a different view of the situation, it was compatible with other agents SA in that each agent's SA formed a composite part of the DSA of the entire network and was required collectively for the entire system to work. The compatible and shared SA views are presented in Figure 5.8.

It is perhaps this finding that is the most important for DSA theory, since it suggests that it may be more pertinent for system designers to strive to design for compatibility of SA, rather than shared SA in multi-agent systems. This means that it may be more pertinent to design systems that provide only the required information to each agent rather than the typical 'all information to all people' approach that is adopted in many collaborative systems. The concept of compatible SA is therefore explored in more detail in Chapters 6, 7, 8 and 9.

The information comprising DSA during both scenarios was also defined and from this it was possible to identify the information elements that each of the agents used

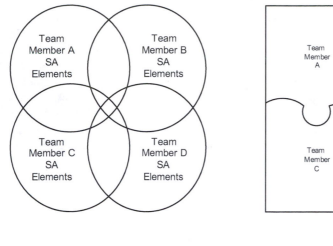

Shared SA (e.g. Endsley &   Compatible SA (e.g.
Jones, 2001)   Stanton et al., 2006)

Figure 5.8   Shared SA versus compatible SA

during the scenarios. For example, during scenario one, it was concluded that the SAPs and APs used the most information elements (77). Again, during scenario two, the APs and the SAPs used the most information elements (91 and 84 respectively) overall. It was concluded that this was a consequence of their involvement in the actual conduct of the switching operations in the field, which meant that they needed to know and understand specific information elements related to the different component task steps involved (e.g. check, open, lock and caution tasks). The COCR operator did not use as many information elements as the SAPs and APs in either scenario (this may seem surprising considering his role in facilitating DSA) since he was effectively supervising or overseeing the activity and did not need to know specific pieces of information pertaining to the tasks required at the remote locations. Rather, the COCR operator appeared to have a high-level overall awareness of the situation (requiring less specific information elements), whereas the SAP and AP had more of a fine-grained, component level awareness of their ongoing task, which required comprehension of more specific information elements.

Information element usage per scenario phase was also defined. For example, during scenario two the COCR operator used the most information elements during phases 2, 3 and 5, whilst the SAP and AP used the most during phases and 4, and 6, and the AP and CP used the most during phase 7. It was concluded that this was a function of their respective roles during each scenario phase, with the COCR operator issuing instructions and permits during phases 2 and 5, and the SAPs and APs performing the operations in the field during phases 4, 6 and 7. These figures are also corroborated by the operational loading figures (from the overall EAST analysis), which indicate that the COCR had the highest workload in terms of operations performed during phases 2, 3 and 5, whilst the SAPs and APs had the highest during phases 4, 6 and 7.

The key information elements extracted from the propositional networks can also be used to make judgements on the system design in terms of the artefacts used and the procedures adopted. Since it is taken that the key information elements represent the most pertinent information related to DSA during task performance, the system in question should ensure that this information is made explicit and that communication of this information is facilitated during task performance. In this case, it was concluded that the key information elements (or their status) identified in both scenarios were all presented explicitly to the agents involved via the artefacts used, the procedures followed or by communications with one another. To give some examples, the information element *work instructions* was distributed around the network (enforced by procedure) by the COCR commander in the form of work instructions and was also read back to the COCR to ensure correct communication, understanding and acceptance. The work instructions were also written down by all parties (as required by procedures). The *operations* information element was distributed around the system by work updates (either given to or requested by the COCR) and was held in the *switching log*, which was maintained by the COCR. Graphical displays also presented some of the key information elements, such as *system state* and *time*, whilst other information elements were in fact physical tools or objects used by the agents during task performance (e.g. *caution tape, cabinets, motor fuses* etc.).

It is therefore concluded that DSA effectively couples distributed systems, in that the information comprising DSA links remotely located agents and structures the communications between them. This mirrors the notion of Stanton et al. (2006) that SA holds loosely coupled systems together. SA in collaborative systems can therefore be defined as each team member's dynamic sharing and usage of systemically held task-related knowledge in order to develop and maintain a compatible and timely awareness of the ongoing situation. DSA refers to the systems overall awareness comprising each of its component agents' compatible SA.

Implications for Collaborative Systems

Perhaps the main implication for collaborative system design to emerge from this study (albeit one that will be further explored throughout this book) is the notion that DSA within collaborative systems comprises each agent's compatible, rather than shared, view of the situation. This is important as it has key implications for the design of collaborative systems and the procedures adopted within them. The shared SA view (e.g. Endsley and Robertson, 2000) suggests that collaborative systems should be designed so that each agent has access to all the information within the system so that SA can be effectively shared across system members. Displays such as common operational picture (COP) displays are used to ensure that team members have a common picture; systems are designed to ensure that each agent's awareness of shared SA elements is identical. The compatible SA view, on the other hand, could suggest that it may be more pertinent to provide each agent with only the information that they require for their specific role and tasks. If each agent's SA is compatible, then why do they need to be presented with information that only other agents working within the system need to know? This is an interesting concept that requires further exploration.

The conclusions from this chapter also suggest that measures can be taken to enhance DSA in collaborative systems. For example, both the theory and the case study evidence suggest that communication and communication links are the key element involved in the acquisition and maintenance of DSA. This follows on from Stanton et al.'s (2006) conclusion that the links between agents in a network are more crucial than the agents themselves in maintaining DSA. Presumably, the information elements required for DSA are, in some form or another, residing within a particular system and so it is communications and the communications links, in addition to this information, that is key to DSA. Take, for example, a system whereby each agent holds information that is critical to the DSA of the entire system, but also requires information from other agents for their specific SA. It is communication between the agents that transmits this information and thus maintains DSA. Further, take the example of a system where the information required for DSA is unknown or missing. Only through communication can this information be located or identified and then dispersed throughout the system. Finally, take the system where information is transmitted from sensors in the field to a central commander at a control base. Should communication fail or be erroneous in this case, the DSA of the entire system may fail. For example, Gorman et al. (2006) describe the Gulf War US Army Black Hawk helicopter tragedy in which 26 people died when two US Army Black Hawk helicopters were mistakenly shot down by two USAF F-15Cs performing routine sweep operations. Gorman et al. (2006) point out that, despite there being various multi-level factors involved, it appears that the tragedy could have been avoided if appropriate communications channels had been exploited. This suggests that in some cases, SA failures may in fact not be failures in SA acquisition, but failures in communication. That is, an individual may have appropriate SA based on what information is communicated to them, since they have perceived and comprehended the information that has been presented to them by the system or by team members. The SA is only erroneous or inadequate because the required information has not been given to the individual. Thus it becomes a communications failure rather than erroneous SA. It seems that communication is the key component to making DSA work. It follows, then, that network links (between agents) are critical. According to Stanton et al. (2006), knowing which links to use (and where to offer information when needed) will determine the quality of DSA.

The content of the propositional networks can also be used to inform system and procedural design. Firstly, it is possible to identify what information is required, when, and by whom, to facilitate effective task performance. Since propositional networks depict what information needs to be known and by whom, it is possible to identify instances in systems whereby the information is either not available (i.e. not presented by interfaces or communication not enforced by procedures) or its dissemination is not supported (i.e. the information cannot be communicated to the people who require it). It is also possible to determine the links between the different information elements underlying DSA, which can be used to design systems and procedures so that linked classes of information are presented or communicated together. Secondly, it is also possible, using social network analysis metrics, to identify the key information elements within a particular network. This is useful since it can be used to strengthen or increase the communication channels that are used to disseminate key information elements within a particular system, or introduce new interfaces that present this information more explicitly. Thirdly and finally, propositional networks can be used to identify what the consequences of removing pieces of information from a system will be. It may be that systems providing certain information elements fail during task performance, and so it is pertinent to see if a system's DSA is sufficient to support task performance when information elements are 'missing'.

It is acknowledged that propositional network methodology employed during this study had a number of limitations. Firstly, unlike when using existing approaches such as the SAGAT (Endsley, 1995b), we were unable to provide a quantitative assessment of the quality of the system's DSA and the SA of the agents working within the system. Therefore, judgements on the quality of the systems DSA were made based on task performance and SME and analyst subjective judgement. Secondly, the use of interview response, verbal transcript and observer data to identify the key information elements used could be criticised for its inability to identify the *tacit* SA-related knowledge (i.e. knowledge used but not openly expressed). However from an analysis of the standard operating procedures for the scenarios analysed, it appears that the propositional networks in this case were comprehensive. It is also conceivable that the links and propositions between the information elements (i.e. relationships between concepts) represent the tacit knowledge used by the operators involved. Thirdly, the data used to construct the propositional networks were subjective and so could potentially be construed as being either prone to error or lacking content. However, the level of subject matter expert input into both data sets reduces the potential for inaccurate data in this case. Finally, the CDM data was collected post-task performance and so could potentially suffer from the various problems associated with post-trial data collection, such as memory degradation (Klein and Armstrong, 2004).

The main aims of this chapter were to investigate further the concept of DSA in complex collaborative environments and to test and extend the propositional network methodology. This study suggests that viewing SA as a systems level emergent property is fruitful for a number of reasons, including that it permits a systemic description of the information comprising SA (which can be extrapolated to an individual SA level) and that it allows judgements to be made on potential barriers to SA acquisition and maintenance. Further, considering SA in this way ensures that team SA within complex collaborative systems is viewed in its entirety, rather than as its component parts (i.e. individual team member SA). In such systems, tasks are rarely performed entirely independently of others, especially in complex situations and when critical decision-making is required (Artman and Garbis, 1998) – these activities tend to require coordinated activity between several individuals (Cannon-Bowers and Salas, 1990; cited in Salas et al., 1995). It is important therefore that SA assessments in collaborative systems consider this coordination.

In closing, it is apparent that further study of DSA during real world collaborative tasks is required. The concept of compatible SA is of particular relevance and further investigation is required into the notion that compatible SA may represent a more appropriate description of how SA works in teams than the shared SA view. Also, work concerning further validation and extension of the DSA theory and propositional network methodology is required.

Chapter 6

Distributed Situation Awareness and Network Enabled Capability Systems: MultiNational Experiment 4

Introduction

The focus now moves to DSA in military systems. Within military command and control systems, SA is a critical commodity (Artman, 2000; Riley et al., 2006) and is often a key factor that distinguishes between mission success and failure. The nature of military systems is such that studying DSA within them is likely to yield significant findings related to the advancement of DSA theory and measurement. Military systems are intrinsically complex and typically feature large teams of agents dispersed over large distances (sometimes even continents) working collaboratively in pursuit of common goals. Tasks are typically performed in complex, rapidly changing, and uncertain settings under time pressure and high-risk levels (Riley et al., 2006). To complicate things further, the military are typically working alongside coalition forces from other nations and other non-military groups such as non-government organisations (NGOs) and charities. Further, the groups are working against an enemy of some sort whose main goal is to defeat them.

Due to continuing technological advances, warfare systems are currently evolving at a rapid rate and therefore the need for guidance on system and procedure design within the military domain is currently apparent. Revolutions in capability, technology and adversaries have influenced the ways in which modern day conflicts are fought and the processes, procedures and tools used are evolving dramatically. For example, coalitions now have a new approach to undertaking war, crisis and peace time operations, known as effects based operations (EBOs). This involves the pursuit of end-states and effects rather than specific actions and involves the consideration of not only military endeavours but also the diplomatic, informational, military and economic (DIME) effects that might influence friends, enemies and neutrals. To support modern day warfare, militaries also have the provision of new advanced technological systems, such as digitised collaborative planning tools and network centric warfare (NCW) or network enabled capability (NEC) supported command and control systems (hereafter referred to as NEC systems). Both (new processes and new technologies) have serious implications for the way in which coalition war and operations other than war (e.g. crisis, peacekeeping and humanitarian aid operations) are conducted and therefore require scientific testing for their effects on operational performance.

The development and implementation of these new processes and technologies makes the concept of DSA critical, both in terms of the design of new military systems, technologies, processes and training programs and for performance evaluation in these new systems to assess how well they are working and the impact that they have on typical operations. The content of DSA and the ways in which these new systems and processes affect DSA requires understanding, as does how best EBO procedures and NEC systems should be designed in order to facilitate DSA.

The aim of this chapter is to investigate the impact that a new NEC-based technology and a new EBO process has on DSA during collaborative military activities. The findings derived from an exploratory study on DSA during the MultiNational 4 (MNE4) experiment, which involved a multi national trial of a new EBO-based concept of operations (CONOPS) using a new digitised collaborative system, are presented. As part of a wider HF analysis the authors collected DSA-related data and used the propositional network approach to assess DSA during the course of the experiment.

Network Enabled Capability

NEC is a currently popular organisational paradigm that involves the use of advanced technology to enhance decision making during operations (Bolia, 2005). As the name suggests, NEC involves the use of linked technological artefacts to enhance information sharing and interaction between elements of warfare systems. As part of the movement towards NEC, militaries worldwide are currently engaged in the process of digitising warfare systems.

The Ministry of Defences Joint Services Publication (JSP) 777 (MOD, 2005) defines NEC as:

> the coherent integration of sensors, decision-makers, weapons systems and support capabilities to achieve the desired effect. It will enable us to operate more effectively in the future strategic environment through the more efficient sharing and exploitation of information within the UK Armed Forces and with our coalition partners. The bottom line is that it will mean better-informed decisions and more timely actions leading to more precise effects.

The supposed tenets of NCW, the US version of the NEC concept, are presented in Figure 6.1.

Much has been said on the enhanced capability that future NEC-based systems are likely to provide, particularly in relation to the enhancement of team SA, and key to the concept is the projected increase in battlespace awareness that they will bring. If we are to believe the hype, the ability of these systems to link distributed forces will lead to increased and enhanced information sharing, which in turn will lead to enhanced SA and team SA, collaboration, self-synchronisation and speed of command. For example, Cebrowski and Garstka (1998; cited in Alberts et al., 1999) suggest that NCW 'focuses on the combat power that can be generated from the effective linking or networking of the warfighting enterprise. It is characterised by the ability of geographically dispersed forces to create a high level of shared battlespace awareness that can be exploited via self-synchronisation and other network-centric operations to achieve commander's intent'

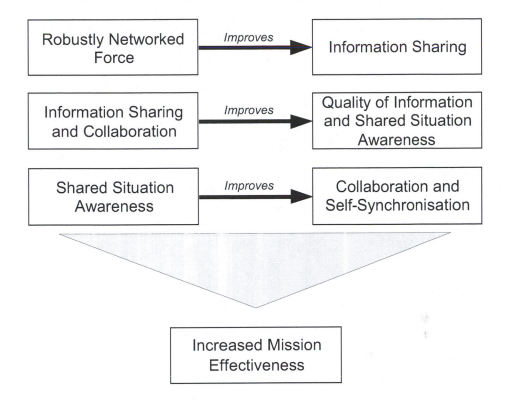

Figure 6.1   Tenets of NCW

Source: Adapted from Alberts and Hayes, 2006.

(Cebrowski and Garstka, 1998; cited in Alberts et al., 1999). The MOD (2005) suggests that 'in the operational environment, it (NEC) will enable shared SA and distributive collaborative working. Alberts et al. (1999) suggest that NEC systems will allow forces to become more knowledgeable due to shared awareness of the battlespace and of the commander's intent, which will in turn allow them to self-synchronise, operate with a smaller footprint and operate autonomously.

On the face of it, then, the concept of NEC seems to bode well for DSA. Distributed force elements should be able to access more information and also share more information and in a more efficient manner. Of course the ability to link every element within a warfare system has great potential for DSA. The importance of communication links in the development and maintenance of appropriate DSA was demonstrated in Chapter 5 and has previously been pointed out by researchers in the area (e.g. Stanton et al., 2006). With the increased potential for linking agents that is brought about by digitised systems, NEC systems have the potential to offer comprehensive communications links within warfare systems, however, when considering SA the information that is communicated around the system, in terms of its content, format, timeliness and accuracy is also of great importance (Alberts et al., 1999; Endsley and Jones, 1997).

The focus of this chapter, then, is not on communications links present but rather on what information is communicated around the system.

Of course, the ability to communicate more information to more people does not necessarily mean that DSA will be enhanced (e.g. Bolia et al., 2007) and simply giving all information to everyone in a collaborative system does not ensure appropriate DSA. Alberts et al. (1999), for example, point out that information has the dimensions of relevance, accuracy and timeliness, all of which are critical to the acquisition and maintenance of appropriate levels of DSA. It is thus important that sufficient consideration is given on how to ensure that each component of a collaborative system receives relevant, accurate and timely information. One of the aims of this study was therefore to investigate what information was available to agents working in the warfare system and also how easily the information that they required could be accessed.

Multi National Experiment 4

MNE4 was undertaken in order to explore concepts and supporting tools for EBOs to assist the development of future processes, organisations and technologies. EBOs is the name given to a recently developed broad framework that supports coalition operations in peace, crisis and war and includes the application of military, political and economic efforts aimed at shaping the behaviour of an adversary (Smith, 2002). EBOs are formally defined as 'co-ordinated sets of actions directed at shaping the behaviour of friends, foes, and neutrals in peace, crisis and war' (Smith, 2002). Effects-based approaches focus on a combination of military and other activities (e.g. diplomatic, information and economic) in order to direct the behaviour of the enemy, friends and neutrals, with the focus being on desired *effects* (end-states) and the *actions* required to achieve these effects rather than merely on actions to be undertaken. According to United States Joint Forces Command (USJFCOM, 2005) the resulting benefits of EBO are a set of actions that are explicitly linked to a set of strategic goals, coherently harmonised with those of other governmental organisations, and made truly adaptive within the course of their execution by effective assessment.

MNE4 involved participants from eight countries (United States, United Kingdom, Australia, Sweden, Germany, Canada, France and Finland) working together as a coalition in a topical warfare scenario. Participants used a new EBO CONOPs to undertake a simulated Afghanistan multinational warfare scenario involving a virtual, distributed, ad hoc, multinational Coalition Task Force (CTF) headquarters (HQ) of 132 participants. The warfare scenario included the functional constructs of knowledge-based development (KBD), effects-based planning (EBP), execution (EBE) and assessment (EBA). Further, supporting constructs such as MultiNational Interagency Group (MNIG) coordination, logistics, information operations, intelligence and medical support were used. Fundamentally, MNE4 involved testing and evaluating a new process of working, the EBOs CONOPS, using new forms of organisation and technology (Information Workspace Collaborative system).

Methodology

Design

The study involved the conduct of an observational study of the MNE4 experiment. The independent variables included the new EBO CONOPS and the information work space (IWS) collaborative system. The dependent variables included DSA, communications (voice and text), meetings, briefings and documents. The hypothesis to be tested was that the IWS system and the KBD group function would enhance the DSA of the system during the activities observed.

Participants

A total of 132 military and civilian participants from the eight countries involved took part in the experiment. Participants were located at headquarters within in each of the eight countries involved and were divided into the following functional groups:

- command group (CG);
- effects-based planning (EBP);
- effects-based execution (EBE);
- effects-based assessment (EBA);
- knowledge-based development (KBD);
- knowledge support (KS);
- knowledge management (KM);
- Multinational Inter-Agency Group (MNIG); and
- components (Grey Cell).

Due to access restrictions and the nature of the study (participants dispersed across eight different countries), it was not possible to collect demographic data for the participants involved.

Materials

Each of the countries involved conducted the experiment within secure laboratories and all information was shared over the Combined Federated Battle Lab Network (CFBLNet). Participants worked in their own countries via telecommunication systems over secure computerised networks; they interacted with one another and undertook daily activities (e.g. planning, meetings, after action reviews etc.) on the IWS collaborative system, which is a collaborative environment comprising text chat rooms, email, radio networks, information databases and Microsoft office tools (e.g. PowerPoint, Word etc.). Participants also had access to the EBO CONOPS, the internet, printing facilities and telephones.

The analysts involved each had access to a desktop PC connected to the IWS system and headphones connected to the IWS system in order to listen to voice communications over the network. The data collection materials used included pen, paper, and CDM questionnaire probes.

Procedure

The participants were presented with a hypothetical modern day Afghanistan coalition warfare scenario. They were instructed to use the EBO CONOPS to support achievement of the desired end state of 'establish a secure environment within the Afghanistan area of operations free of internal and external threat'. The main goals of the coalition included establishing a regional representative government of Afghanistan, establishing rule of law, pursuing humanitarian national development, building the capacity of the Afghanistan security forces and establishing conditions for enhanced opportunities for legal economic and social development (USJFCOM, 2005).

The experiment consisted of three, five-day periods (Monday to Friday) and took place over a three-week period between 27 February and 17 March 2006. Daily activities began at 9.30am and finished at 7.30pm. A daily battle rhythm was used to guide the experiment, and scenario injects were used to manipulate the scenario faced by the participants.

The analysts were located at the UK HQ during the experiment and observed activity taking place in the UK HQ and on the IWS collaborative planning system. This gave the analysts access to voice communications and text chat taking place between all participants, and the meetings taking place between participants and all of the documents used. Due to the size of the experiment and breadth of the activities undertaken, the analysts focussed on small vignettes as opposed to the entire experimental scenario; this approach is advocated by the NATO code of best practice (NATO, 2002) for studying command and control systems, suggesting that an appropriate way of dealing with the complexity of such systems is for analysts to present their findings in the form of vignettes. The analysts therefore concentrated on meetings and planning sessions and recorded information on what was happening during a particular vignette, including the communications between participants, the content of the communications, the topics being discussed and any documents referred to. CDM interviews were also conducted with participants where possible. The primary sources of information used by analysts were the CONOPS, the commander's briefs, observation of meetings, communications between participants (voice and text chat) and any documents referred to during the meetings and briefs.

The data collected was used to construct propositional networks for each vignette observed during the experiment. The analysis focused specifically on the activities of the following five functional groups:

1. *KBD*. The KBD group was responsible for building and maintaining a knowledge base that participants could use for developing SA and understanding, including

SA on the operating environment, the adversary, and friendly and neutral forces.

2. *EBP*. The EBP group was responsible for developing and refining operational plans, using DIME planning, that establish clear links between the commanders desired end state and the effects required to achieve the end states.

3. *EBE*. The EBE group was responsible for coordinating, directing and monitoring task force operations.

4. *EBA*. The EBA group was responsible for assessing the actions and effects in order to identify operational deficiencies and recommend methods to improve force effectiveness.

5. *MNIG*. The MNIG was responsible for the civilian component of the response. This involved harmonising the planning and actions of civilian agencies and the coalition task force, coordinating provision of civilian capabilities, expertise and perspectives, enabling collaboration with coalition partner agencies, international organisations and NGOs and facilitating information sharing amongst coalition government agencies, militaries, international organisations and non-governmental organisations.

Figure 6.2 presents a summary of the role and main activities of the groups analysed.

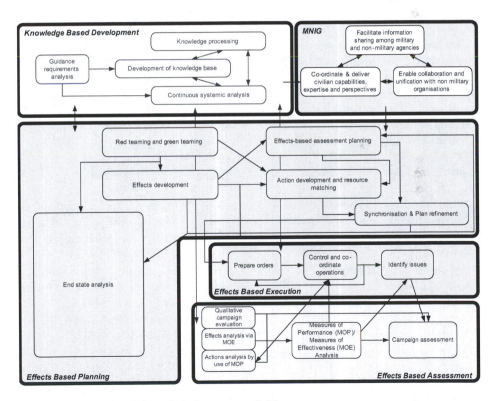

Figure 6.2   MNE 4 functional group activities

Results and Discussion

Functional Group Distributed Situation Awareness Requirements

CDM interviews were conducted, using the probes presented in Chapter 4, with an SME from each of the following groups: KBD, EBP, EBE and EBA. From the CDM data, high-level propositional networks were constructed for each group in order to identify the DSA requirements of the different groups during EBO-based activities. The resultant propositional networks are presented in Figure 6.3.

The propositional networks presented in Figure 6.3 demonstrate that, due to their distinct roles in the EBO process, the groups involved had very different DSA requirements during the activities undertaken. The KBD group's role was to provide the knowledge base required for the other groups to acquire and maintain an appropriate level of DSA throughout operations undertaken, which meant that their DSA requirements were related to collecting and representing information related to what the other groups needed to know. Their DSA requirements thus included an understanding of what it was that other groups needed to know and also an appreciation of this information itself, including applicable policies, laws and strategies, intelligence products, plans and guidance, the CONOPs and the activities of enemy, friendly and neutral forces. The EBP groups DSA requirements were primarily concerned with the development of effects and plans and included the commander's intent, effects and desired end state, resources and the current situation. The EBE group, whose role was to coordinate, direct and monitor task force operations, had DSA requirements related mainly to the activities on the ground, including information regarding the activities being undertaken by the components and other groups. Finally, the EBA group was concerned with the progress of the plans being enacted and subsequently their DSA requirements mainly comprise measures of performance and effectiveness and the progress of activities being undertaken.

Commanders Brief Propositional Networks

Overall scenario propositional networks were constructed from the commander's briefs that were given at the start of each day. The commander's brief described all activities from the previous day, forthcoming activities and included situation reports from each of the different groups (e.g. KBD, EBP, EBE, EBP, EBA and MNIG). The purpose of constructing the propositional networks was to identify specifically what information was being used by the different groups involved and what information was being shared between groups. Due to size restrictions, the entire propositional networks cannot be presented and only extracts are used. Extracts from the propositional networks constructed from the commander's briefs on 3 and 9 March are presented in Figure 6.4.

The usage of information elements by the different groups involved was calculated based on each group's reference to the topics during their daily situation reports. The overall information element usage for the 3 and 9 March commander's brief propositional networks is presented in Figure 6.5.

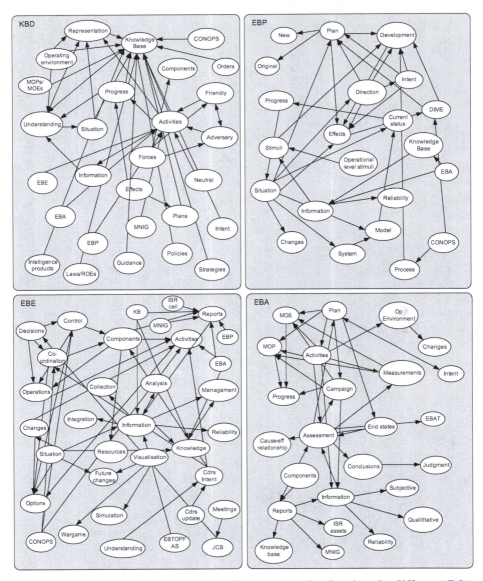

Figure 6.3   Functional group propositional networks showing the different DSA requirements of the KBD, EBP, EBE and EBA groups

The propositional network analysis of the commander's briefs indicated that, initially, the information contained within the knowledge base was not being accessed and used by the other groups involved. As the observation progressed, however, the usage of information by the different groups and the communication of information between the KBD and the different groups increased. For example, initially (i.e. 3 March network) the KBD group held the majority of the information related to the scenario and this information, although available on the IWS, was not yet being used

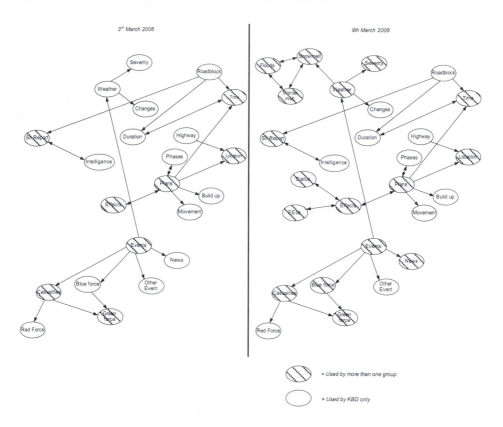

Figure 6.4   Commander's brief propositional network extracts

Note: Figure shows how information usage increased over the course of the experiment.

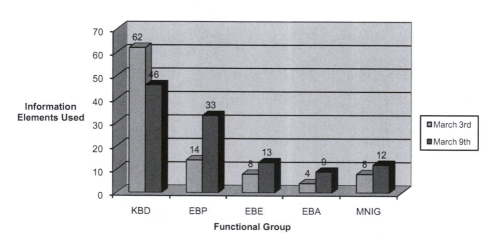

Figure 6.5   Information element usage

by the other groups involved. This suggests that at this point in the scenario, the other groups' DSA (and overall DSA of the system) was degraded somewhat, since groups were not using the information as required. The 9 March propositional network extract, however, demonstrates how the usage of the information provided by the KBD group had increased significantly. For example, in the 3 March propositional network, a total of 40 information elements held by the KBD group remain unused by the other groups involved. In the 9 March propositional network, the total number of unused information elements decreased to 21. This suggests that the groups' and the overall system's DSA had improved somewhat, since they were now using more of the information required to support task performance.

The propositional network analysis also indicated that a commonality in the information being used by the different groups also increased. For example, on 3 March, the total number of information elements being used by three or more of the groups was two. This figure had increased to 20 by 9 March. This was taken to show greater consensus on the scenario and activities required and a commonality on themes between groups.

The initial lack of usage of information held by the KBD has interesting ramifications for technologically supported and distributed multinational warfare operations. The provision of this information on the IWS by the KBD group indicates that the information was available to (and to some extent required by) other groups. The non-usage of the information by other groups suggests that they were not accessing the information as required. Further, when the groups proactively went looking for the information that they required our observations suggest that they either could not locate it on the IWS or did not know who to speak to within the organisation to find the information. The observational study also indicated that the KBD group was waiting for information requests (rather than proactively contacting the groups). During the experiment there was some confusion over the KBD group's role, with some participants viewing them as 'signposters' (i.e. pointing participants to information) and others viewing them as gatekeepers (i.e. supplying information on demand). The CONOPS suggested that the KBD group's role was as gatekeeper, supplying information to other participants as and when requested.

This finding is particularly significant for NEC-based systems since it seems to suggest that, if not designed appropriately, DSA may be initially inhibited by new NEC technology and process. The results demonstrate that the system's DSA was inadequate initially and that the sub-teams within the system were not aware of who knew what or of where to find the critical information that they required in order to undertake their activities. This suggests that the enhanced connectedness of an organisation and the presence of a knowledge base do not guarantee enhanced sharing of information and DSA, as is argued by proponents of such systems.

Humanitarian Aid Propositional Networks

A critical humanitarian aid scenario unfolded over the course of the experiment. To clarify, humanitarian aid 'includes programs conducted to relieve or reduce the results of natural or man-made disasters or other endemic conditions such as human pain, disease, hunger, or privation that might present a serious threat to life or result in great damage or loss of property' (US Army, 1994). Humanitarian aid operations are particularly relevant when discussing DSA since it involves many stakeholders (including the Army, national government organisations, private organisations, international organisations and charities, to name only a few) working together towards a common goal. Propositional networks related to the humanitarian aid situation were constructed from the daily updates and situation reports made during the experiment.

Examples of the humanitarian aid propositional networks are presented in Figures 6.6 to 6.9. Information elements with bold lines represent those that were being used at that point in the scenario. Information elements without bold lines represent those information elements that were no longer being used at that point in time in the scenario. The initial propositional network from 6 March (Figure 6.6) suggests that the different participants held a very high-level awareness of the humanitarian aid situation at that

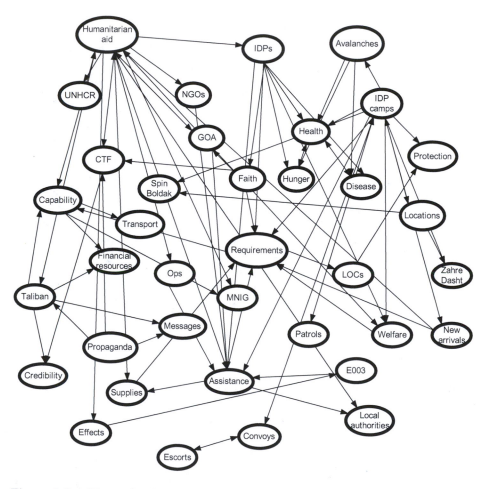

Figure 6.6   Humanitarian aid operations propositional network 6 March

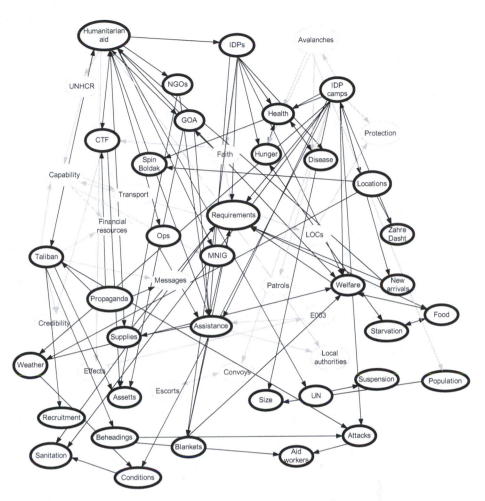

Figure 6.7   Humanitarian aid operations propositional network 7 March

time, including a basic knowledge of the overall humanitarian aid requirements and of the different locations of the internationally displaced persons (IDP) camps. They had a general awareness of what was going on but did not have detailed knowledge related to the situation and its implications and requirements. Significantly, much of the information required to enhance their awareness of the situation was available on the IWS collaborative system, but was not accessed at this time.

As the observation progressed, the humanitarian aid situation deteriorated, which led to an increase in the associated information requirements (i.e. information required by participants in order to deal with the situation accordingly). This also meant that the participants required more detailed SA of the situation in order to deal with it accordingly. Correspondingly, the KBD group began to make more information available to other participants within the system, which led to the different groups being able to develop a very detailed, specific SA of the situation. For example, the propositional network

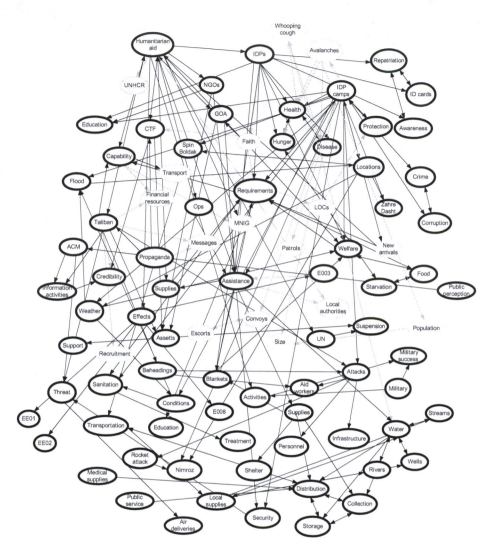

Figure 6.8   Humanitarian aid operations propositional network 14 March

constructed on 15 March (Figure 6.9) shows that groups had developed a detailed level of SA related to requirements and events in the camps. To demonstrate, rather than simply knowing that the IDPs need water, the 15 March propositional networks shows how the groups are now considering where the water will come from (e.g. 'supply' and 'local suppliers' information element), how the water will be treated (e.g. 'treatment') and distributed (e.g. 'distribution', 'transport' and 'routes') and how it will be stored (e.g. 'storage' and 'wells'). The expansion of information elements related to the water shortage is presented in Figure 6.10.

The expanding propositional networks presented in Figures 6.6–6.9 therefore demonstrate the increase in SA requirements and usage of information related to the

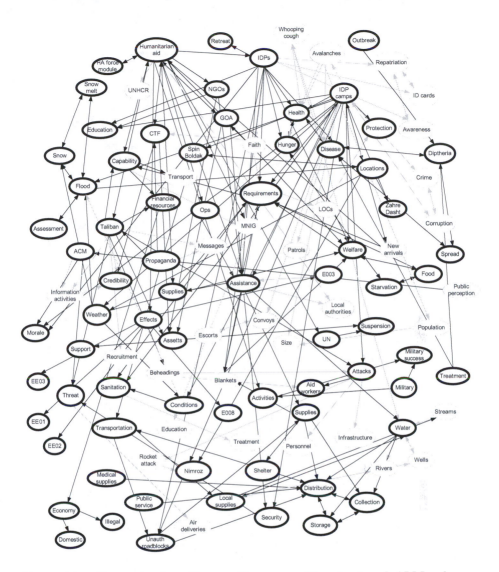

Figure 6.9   Humanitarian aid operations propositional network 15 March

humanitarian aid situation over the course of the experiment and thus demonstrate the dynamic nature of DSA in collaborative military environments. As the situation unfolded, SA requirements increased and thus more and more information was required to maintain an adequate level of SA for dealing with the situation. The different participants and groups needed to know more about the situation in order to deal with it. Further, the increased amount of information elements depicted in the propositional networks demonstrates how more information related to the humanitarian aid situation was made available and used by different groups within the organisation.

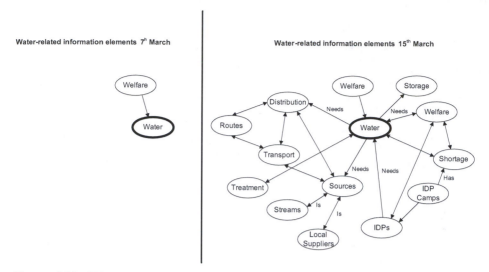

Figure 6.10   Water-related SA information elements

The key information elements were also extracted from the humanitarian aid scenario using sociometric status and centrality statistical calculations (see Chapter 5 for descriptions of each metric). Key information elements are defined as those that have salience for each scenario phase, salience being defined as those information elements that act as hubs to other knowledge elements. Those information elements with a sociometric status value above the mean sociometric status value and a centrality score above the mean centrality value were identified as key information elements. The key information elements for the humanitarian aid scenario are presented in Table 6.1.

As demonstrated in Table 6.1, the key information elements at the beginning of the scenario (6 and 7 March) as the humanitarian aid situation began to unfold were the overall *humanitarian aid situation*, the *IDPs*, the *IDP camps* and the *humanitarian aid requirements*. This reinforces the notion that the participants at this time held only a very high level awareness of the situation. It is important to reiterate, however, that at this point some of the information required for a more detailed level of SA on the situation was held by the system (i.e. the KBD group) but, due to various reasons which will be discussed later, the different components of the system did not access this information. This is a key finding that has direct relevance for the design of NEC-type systems and processes, that is, how do system and procedural designers ensure that the information that is held by the system percolates to the appropriate system elements in a timely manner?

As the situation began to deteriorate, the key information elements began to increase in line with the requirements and various events occurring within the IDP camps. Factors such as *food, water, sanitation, supplies* and *distribution* began to become of primary concern, along with the *Taliban*, who were beginning to infiltrate IDP camps, *attacks* on IDP camps and United Nations (UN) workers within the IDP camps, the *effects* being planned, the *coalition task force* and *repatriation* efforts and the *ACM* and *Taliban propaganda*. It is significant to note here that along with the increased

Table 6.1   Key information elements related to the humanitarian aid scenario

6th March	7th March	14th March		15th March	
Humanitarian Aid	Humanitarian Aid	Humanitarian Aid	Food	Humanitarian Aid	Food
Internationally Displaced Persons	Internationally Displaced Persons	Internationally Displaced Persons	Water	Internationally Displaced Persons	Water
Internationally Displaced Person Camps	Internationally Displaced Person Camps	Internationally Displaced Person Camps	Sanitation	Internationally Displaced Person Camps	Sanitation
Requirements	Requirements	Requirements	Supplies	Requirements	Supplies
	Taliban	Taliban	Distribution	Taliban	Distribution
		Locations	Attacks	Locations	Attacks
		Operations	Threat	Operations	Threat
		Coalition Task Force	Repatriation	Coalition Task Force	Repatriation
			Effects	ACM	Effects
					Propagenda

information requirements, the participants were now becoming more proficient with the IWS collaborative system and CONOPs procedure and were also more aware of the specific roles of the different groups involved. This meant that the different players involved were effectively aware of more information, as they knew both where to find it and who would know what within the system. The key information elements were all used by the different participants and groups involved.

Joint Planning Group Targeting Meeting

To demonstrate the outputs derived from the meetings observed, a propositional network developed from a Joint Planning Group target prioritisation meeting is presented in Figure 6.11. The meeting itself involved discussing and prioritising the engagement of potential targets. The agents involved in the meeting were distributed and communications between agents were made using the IWS system, which allowed both verbal (using microphone headsets) and text (using conventional keyboard) chat.

Figure 6.11 provides an example of how, within meetings taking place over the course of the experiment, there was some difficulty in addressing non-military effects, e.g. there were problems with fitting diplomatic, informational and economic effects

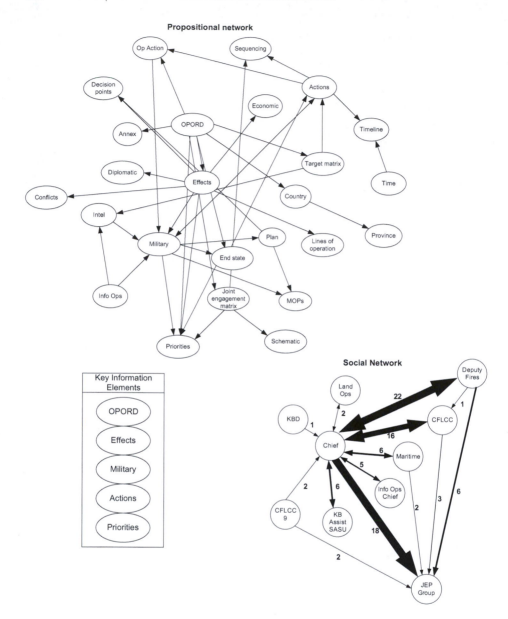

Figure 6.11 Joint engagement planning group target prioritisation meeting propositional network, key information elements and social network diagram

into a military context. In this case, although the diplomatic and economic effects were raised initially, the discussion was very much focused on the military effect of destroying the targets being discussed and MNIG representatives found it difficult to raise non-military issues associated with the effects being sought. This is demonstrated within the propositional network by the linkage between the 'military' information element to the other elements 'effects', 'OPORD', 'info ops', 'intel', 'actions', 'op actions', 'priorities', 'MOPs' and 'plan', whereas the other non-military effects, diplomatic and economic, are linked only to the 'effects' nodes. This demonstrates how the meeting was focused primarily on planning the military effects, including the actions required, the measures of performance and resultant orders, whereas the other diplomatic and economic aspects were only briefly considered. The 'military' information element was also one of the key information elements identified, as shown in Figure 6.11.

Also demonstrated in this meeting was the confusion over the KBD groups role within the EBO process; the meeting chair had the expectation that KBD representatives would supply information on request, whilst the KBD representatives saw their role as signposting information, telling other participants where to find it. Within Figure 6.11, the social network diagram shows that the KBD representative was not engaged in communications with the other parties, which suggests that they were not supplying information as required.

Finally, it was also observed during this meeting that there was a great deal of reference to the CONOPS for clarification purposes, which suggested that participants were still learning their roles and responsibilities within the EBO process. This is demonstrated within Figure 6.11 by the high number of incoming communications to the chief, a significant number of which were requests for clarification.

Conclusions

The aim of this chapter was to investigate the concept of DSA in a future military multinational operational environment. Specifically, the study focused on the effects that the new technology (IWS collaborative system) and process (EBO CONOPS) had on the content and structure of DSA during the experiment. In relation to this research, the study was also used as a means of further investigating the concept of DSA and its measurement. A summary of the main findings is presented below.

Distributed Situation Awareness

The propositional networks presented demonstrate that, as the experiment unfolded, more information was used by, and passed between, the different groups involved. The findings also indicated, however, that participants initially did not access and use the information required for SA that was held by the KBD group; although the system held the information required for SA, it was not initially utilised. It is concluded that this was a result of two factors. Firstly, confusion over the KBD's role meant that the KBD group was not initially supplying groups with information; rather they were waiting for information requests, which caused a delay in information being disseminated as required. Secondly, some of the participants were not sufficiently proficient with the IWS system (technology) and the CONOPS (process) and thus found it difficult to locate the information that they required. They did not know either where to look to find information or who to speak to in order to get it. Both system shortcomings improved as the observation progressed, which is demonstrated by the increase in size of the propositional networks presented in Figures 6.6–6.10, which indicates that participants were gaining access to and using more information towards the end of the experiment. It was notable, however, that information tended to be discovered rather than supplied. That is, participants had to find the information that they required rather than being supplied with it by the KBD group. This was incongruent with the CONOPS, which outlined the KBD group's role as an information supplier to the other groups. Further, the provenance of information was also often unknown. There was often no indication of the source of the information within a document (i.e. it was not possible to trace it back through the EBO process) nor was there any indication of how appropriate the information was in terms of quality, relevance and age. This had an impact on the confidence in either the quality of the information or its time-sensitivity, which is critical given the high tempo nature of EBO.

The quality of DSA improved as the observation progressed. Initially participants held only a very high level SA of the ongoing situation; in fact it could be argued that group SA was initially in a diminished state. As an example, the initial humanitarian aid propositional networks demonstrate that, although participants were aware of the situation, they did not possess a detailed awareness of the situation in terms of what the problems were, what was required and what was to be done. Instead, they were aware that there was a humanitarian aid situation unfolding but their knowledge of specific problems and requirements in the camps was limited. As the observation progressed, however, player SA developed to a more fine-grained level. It was concluded that this was due a clarification of roles that occurred as the observation progressed and also to the participants becoming more proficient with the EBO process and the IWS technology (i.e. they were getting better at finding information on the system).

Again, the concept of compatible, rather than shared, SA was demonstrated throughout the activities analysed. In this case, the compatibility of SA arose more out of the different roles of the different groups involved over and above anything else. Propositional networks developed for each of the main functional groups involved (e.g. KBD, EBP, EBE and EBA) from CDM data demonstrated that each group had very different DSA requirements during the EBOs observed. The KBD group's SA, whose responsibility it was to provide the knowledge base for other groups' SA, consisted of an awareness of what other groups needed to know and also what information was available and where it was available from. They effectively took on the role of DSA facilitators since they were told what groups needed to know and built the knowledge base accordingly. The EBP group was concerned with developing and refining operational plans and so their SA concerned the current operational situation, the desired end state, resources available and what was required. Their SA was planning orientated. EBE, whose role it was to coordinate and monitor task force operations, held SA of the ongoing activities on the ground, i.e. the status of the execution activities. EBA, on the other hand, had to assess actions and effects to identify any operational deficiencies and so their SA comprised execution activities on the ground and also potential approaches to improve force effectiveness. Finally, the MNIG group, who were concerned with the civilian component of the response, held SA relating to non-military aspects of the actions and effects being undertaken (i.e. the civilian role in the plans being executed).

Effects Based Operations

It was concluded that the new EBO CONOPs affected DSA in terms of the information used for DSA and the dissemination of this information throughout the system. It was evident, for example, that there was some confusion between participants as to the exact nature of the roles of the different groups within the EBO approach. For example, the degree of confusion over the KBD's role during the experiment demonstrates this. In MNE4 briefings, the role of the KBD function was discussed in terms of supplying information; however, many participants had the expectation that the KBD's role was to supply information on request, whereas KBD representatives felt that their role was to point participants in the direction of information and tell them where to find it. It was concluded that this confusion was a result of ambiguity within the CONOPS. In this case the CONOPS specified the 'who' for an activity, but not the 'how', allowing contributors to interpret their roles differently. This ambiguity may also have been the reason that parts of EBO seemed to ignore diplomatic, information and economic input. For example, the CONOPS specified that a representative has to be at a particular meeting, but it did not say what role they should play – without training in diplomatic, information and economic elements, they may find it difficult to incorporate these domains into planning and execution. This confusion of roles had an impact on DSA during the course of experiment. During the meetings, there was a great deal of reference to the CONOPS for clarification purposes, which suggested that participants were still 'learning the rules' for the experiment. This led to a degree of uncertainty about what was required and had a bearing on DSA in terms of participants knowing what information they required at different times throughout the experiment. These findings can be related to the findings derived from the energy distribution study presented in the previous chapter, which suggested that clear role definition and appropriate procedures were key aspects which facilitated the high levels of DSA held by the energy distribution system in the scenarios observed.

The MNIG group also found it hard to introduce non-military aspects of effects. DSA remained, in the main, military specific and the diplomatic, political and economic aspects of effects were often ignored.

Implications for the Design of Future Warfare Systems

Whilst these findings again reinforce the conclusions arrived at in the previous chapter that communications links are the critical factor in maintaining a systems DSA (e.g. Stanton et al., 2006; Gorman et al., 2006), they also demonstrate that there is much more to enhancing DSA than fostering the appropriate communications links alone. Alongside the appropriate communications links being present within collaborative systems, the findings also suggest that clarity of role definition, the use of appropriate and clear procedures and the presentation of information in terms of where and to whom it is presented and to whom are also critical issues. In particular, the clear specification of where to find information and/or who to get it from and the supply versus discovery of information are interesting issues raised by this study. Further, the findings suggest that having a repository of information designed to facilitate DSA development and maintenance may not be appropriate if users are not well versed in its usage.

Within multi-agent collaborative systems, roles should be clearly and explicitly defined so that the entire system understands what each component (human and non-human) should be doing and what information they should be contributing to the DSA of the overall system. A lack of clarity in role definition may mean that components do not appreciate who is likely to know what, where information is held and how it can be accessed. A related factor is the clear definition of where and who to get information from within collaborative systems. This allows each component of the system to be able to locate and access the information that they require for their activities quickly. Further, the findings suggest that information should be grouped (or presented) together, based on the relationships between them, in order to support DSA. For example, information regarding water requirements in IDP camps could be presented with links to other related information concerning water sources, water storage and water distribution. Linking information in this way supports DSA development, since the relationships between concepts are being supported by the information presented by the system. Finally, an important issue surrounds whether system components should be supplied with the information that they require or whether they should be left to their own devices to find it themselves. Whilst both aspects may be evident in collaborative systems, it is important that system components are aware of how they are to get information, i.e. if they are required to find information themselves then they should be fully aware that they have to do so.

The findings therefore suggest that the design of multinational warfare procedures and systems should focus not only on providing the appropriate communications links within a network but also that the links available should be made explicit to the agents involved. Agents need to know which links to use in order to access information and also where to offer information when needed during task performance. The findings also suggest that it is important that within collaborative systems, agents know who knows what in order for the system's DSA to function effectively. This 'meta SA' (awareness of other agents 'awareness) is therefore a critical element that is required to support efficient and effective DSA. Finally, the importance of clear role definition within collaborative systems has again been reinforced.

Propositional Networks

Further validation for the propositional network methodology as a means for describing and evaluating DSA during complex collaborative activity was offered by this study. In this case, the propositional networks permitted the description of DSA from the point of view of multiple agents within the system, which in turn allowed the definition of the information used by different agents at different times throughout the scenario. The propositional networks developed in this case were also useful in that they depicted how the information related to DSA expanded and how DSA became richer during the experiment. The water example presented in Figure 6.10 demonstrated how an analysis of the different information elements within the propositional networks allows one to evaluate how rich a system's DSA is at a particular time. This is particularly useful for assessing the quality of the information that is passed around a network and can also be used to identify instances where the key information required for DSA is either not available (i.e. not presented by interfaces) or its dissemination is not supported (i.e. the information cannot be communicated to the people who require it). The usefulness of vignettes, as opposed to whole scenarios, to describe and analyse DSA was also demonstrated by this analysis, which suggests that, in complex, large-scale scenarios it is acceptable to use vignettes rather than the whole scenario (which leads to large and unwieldy outputs).

In closing, it is concluded from the analysis that DSA was affected by both the new process adopted (effects based operations) and the new technology used (e.g. IWS). The new EBO process led to some confusion over the roles of different participants in disseminating information whilst problems with the IWS meant that participants found it difficult to locate and assimilate DSA-related information. It is recommended, therefore, that further study into the effects of new coalition processes and technologies on collaborative SA is undertaken and that measures are taken to design such processes and systems to support, rather than inhibit, SA during multinational operations.

Out with the Old and In with the New: A Comparison of Distributed Situation Awareness Using Analogue and Digital Mission Planning Systems

Introduction

The findings derived from the study presented in the previous chapter indicated that there may be significant issues associated with the design and implementation of future digitised NEC-based warfare systems and the impact that they have on DSA during warfare mission planning and execution activities. This research now moves towards the consideration of how warfare support systems should best be designed to enhance, rather than inhibit, DSA and focuses specifically on a newly developed land warfare digital mission support system that is currently being tested and refined in the UK military domain. This chapter presents a DSA-based analysis of the digital mission support system and a comparison of DSA when using the new digitised system and the old analogue paper map system.

Digital Planning Systems

Much has been said on the enhanced capability that future NEC systems are likely to provide (see previous chapter). Central to the utility of such systems is the use of linked electronic or digitised mission support systems, which will conjugate together to form the 'network' in NEC. The underlying belief of the proponents of such systems is that the provision of larger, better-connected networks will enable more information to be communicated more quickly and to more people, which in turn will enhance team SA and tempo during mission planning and execution. In essence, it is postulated that missions will be planned and executed more quickly, more efficiently, with enhanced SA, tempo and increased levels of collaboration and with access to more information than in previous times.

In line with the military's wholehearted movement towards NEC-based systems, there has been a recent spate of digital mission support systems being developed, tested and even introduced in theatre. In the UK military, for example, the digital mission support system focused on in this chapter and the AeroSystems suite of Mission Planning Systems (MPS) for military aircraft are currently being developed, tested and used in theatre.

It is our opinion that great caution should be taken when forecasting the benefits of NEC-based systems. The overriding assumption that more information will enable better performance is a worrying one when one considers aspects such as distributed information requirements, the format in which information is presented and the amount of data that can be presented versus the amount of data that can be meaningfully processed. Bolia et al. (2007), for example, suggest that the increased amount of information available in such systems does not necessarily mean that users of the data will make better decisions, due to a number of factors, including that increases in quantity of information do not necessarily lead to an increase in the amount of relevant information, the fact that all data is not good data and that false data could be deliberately be fed into networks or data could be erroneous and also that data is only as good as its interpretation.

Designers of mission support systems clearly need to consider how the system can be designed so that it enhances, not inhibits, SA during the activities that it is being designed to support. Endsley and Jones (1997), for example, point out that the way in which information is presented by such systems ultimately influences SA by determining how much information can be acquired, how accurately it can be acquired and to what degree it is compatible with SA needs (Endsley and Jones, 1997).

It is clear that there needs to be great consideration given to the impact that new technologies have on performance during mission planning and execution activities. Military mission planning and execution processes are time served and are efficient and effective as they are, and so the addition of software-based mission support tools should be made with great care. In particular, the impact of new technological systems on DSA during mission planning and execution activities is a key aspect of how well such systems will work and thus requires testing throughout the design lifecycle.

The purpose of this chapter is to present the findings derived from a study, part of which was undertaken in order to investigate the impact of a newly developed digital mission support system on DSA during the land warfare mission planning and execution process. The study focuses on the Combat Estimate (CE) seven questions planning process, which can now be undertaken electronically (as opposed to the traditional, analogue, paper map process) using the newly developed digital mission support system. To do this, case studies on DSA during the old paper map planning process and the new digital system planning process are compared and contrasted in order to identify the impacts, good and bad, on DSA that the new mission support system has had. It is intended that the findings from this analysis will inform the development of guidelines for the design of software-based mission support systems.

The Digital Mission Support System

Within the land warfare domain, the UK Ministry of Defence is currently developing and testing the digital mission support system that this chapter focuses on. The tool is a digitised battle management system that provides command and control support for battlefield planning and execution tasks and has been designed in order to enhance SA, information management, planning, control and tempo during mission planning and execution. According to the system's creator, the tool provides the support for data distribution, planning, collaboration and execution. Amongst the aspirations for the system, one of the key enhancements that it is claimed it will bring to mission planning and execution is enhanced SA. For example, the following quote regarding the system is taken from the Army's website:

> it will provide enhanced situational awareness and common operational, intelligence, personnel and logistic planning tools to improve the tempo, survivability and effectiveness of land forces. It will also facilitate mission analysis and the provision of orders, map overlays and route planning and provide standard reports and returns formats. (MoD, 2007)

One of the processes that the system has been designed to support is the current military land warfare mission planning process, the CE (MoD, 2007). More commonly known as the 'Seven Questions' this involves working through a process of seven separate questions in order to understand the battle area and the enemy's intentions and then develop, select and resource appropriate courses of action. A brief description of the seven questions process is given in the following section.

The 'Seven Questions' Planning Process

The seven questions mission planning process is a collaborative process that consists of the following seven questions:

Question 1 – What is the enemy doing and why?
Question 2 – What have I been told to do and why?
Question 3 – What effects do I want to have on the enemy and what direction must I give to develop my plan?
Question 4 – Where can I best accomplish each action/effect?
Question 5 – What resources do I need to accomplish each action/effect?
Question 6 – When and where do the actions take place in relation to each other?
Question 7 – What control measures do I need to impose?

Each question is undertaken by various cells within the organisation (e.g. Battle Group or Brigade) and the products from each question are used to inform the other questions in the process. A brief description of the process is presented below.

Question 1 involves the use of maps to undertake the battlefield area evaluation, which involves the terrain analysis, threat evaluation and threat integration processes. The terrain analysis component requires an assessment of the effects of the battlespace on enemy and friendly operations and involves the identification of likely mobility corridors, avenues of approach and manoeuvre areas. For the terrain analysis phase, the mnemonic 'OCOKA' is used, which comprises the following aspects of the terrain (MoD, 2007):

- **Observation;**
- **Cover and Concealment;**

- **Obstacles**;
- **Key** terrain; and
- Avenues of Approach.

Other key SA requirements during the terrain analysis include the weather, restricted areas and potential choke points.

The threat evaluation phase involves identifying the enemy's likely *modus operandi* by analysing their tactical doctrine, past operations and their strengths and weaknesses. The end state of the threat evaluation phase is to 'visualise how the enemy normally executes operations and how the actions of the past shape what they are capable of in the current situation' (MoD, 2007, p. 12). Key SA requirements here include the enemy's strength and weaknesses, their organisation and combat effectiveness, equipment and doctrine and also their tactics and preparedness.

The threat integration phase then involves combining the battlefield area evaluation and threat evaluation outputs in order to determine the enemy's intent and how they are likely to operate. The products of the threat integration include the enemy effects schematic, which details their likely course of action, situation overlays for each course of action identified and an event overlay. Key information during this phase includes the Named Areas of Interest (NAIs) and likely enemy courses of action.

The SA requirements during Question 1 are therefore primarily related to the battlefield area itself and also the enemy and their likely *modus operandi*. The output is an understanding of the battlespace and its effects on how the enemy (and friendly forces) are likely to operate. The key SA requirements during Question 1 are therefore related to the terrain (e.g. OCOKA, weather, key terrain etc.), the impact of the terrain on the enemy's likely actions, the enemy's strengths and weakness (including combat effectiveness, resources and capability, doctrine and past operations) and also the enemy's likely courses of action (based on a comparison of the terrain with the enemy's capability and past activities).

Question 2 is known as the mission analysis and asks 'what have I been told to do and why?' Of specific interest during Question 2 are the specified and implied tasks and the freedoms and constraints of the mission. Undertaking the mission analysis involves completing a mission analysis record, which requires a statement of the mission both two up and one up, a statement of the main effort, specification of the specified and implied tasks, their deductions, any requests for information (RFIs), the commander's critical information requirements (CCIRs) and finally the freedoms and constraints associated with the mission. Specified tasks are typically found in the mission statement, the coordination instructions, the Decision Support Overlay (DSO), the Intelligence collection plan and the Combat Service Support Operations (CSSO) (MoD, 2007). The SA requirements extant during Question 2 are therefore the mission itself, the main effort, the resultant specified and implied tasks from the mission and main effort and the freedoms and constraints of the mission.

Question 3 involves the commander specifying the effects that he wishes to have on the enemy (MoD, 2007), what is referred to as his battle winning idea, or 'that battlefield activity or technique which would most directly accomplish the mission' (MoD, 2007, p. 23). Based on the information gleaned from Questions 1 and 2, the commander should now understand the battlespace area and the aims of the friendly forces involved and should comprehend how the enemy are likely to operate. Using this understanding, the commander then identifies the effects required in order to achieve the mission and prevent the enemy from achieving their mission. The commander specifies his effects using an effects schematic and gives the purpose and his direction to the staff for each of the effects described. Additionally the commander also specifies what the main effort is likely to be and also his desired end state. Additional direction designed to focus the planning effort is also given at this stage. This might include guidance on the use of applicable functions in combat, principles of war and principles of the operation (MoD, 2007). Finally, the commander confirms his CCIRs and RFIs.

Questions 4, 5, 6 and 7 are primarily concerned with the development of the courses of action required to achieve the commander's desired end state. Question 4 involves identifying where each of the actions and effects specified by the commander will be best achieved in the present battlespace area and involves placing the commander's effects, NAIs, TAIs and decision points (DPs) on the battlespace. Although some of the effects are likely to be dictated by the commander and the ground, others, such as STRIKE and DEFEAT effects, can often potentially take place in a number of areas depending on a variety of factors such as enemy location, terrain and friendly force capability. The output of Question 4 is the draft DSO which contains the commander's effects, NAIs, TAIs and DPs for the mission. The key SA requirements during Question 4 are therefore the actions and effects specified by the commander (the commander's battle winning idea), the outputs of Questions 1, 2 and 3 (i.e. an understanding of the battlespace and its impact on the enemy and also of the enemy's capability and likely courses of action).

Question 5 involves specifying resources for each of the commander's effects, NAIs, TAIs and DPs. This involves considering the effects required and then the mission, combat power, type, size and strength of the enemy at each NAI and TAI. Much of this information can be derived from the assessment of the enemy's strengths and weakness made during Question 1 as part of the threat evaluation. The output of Question 5 is a series of potential courses of action for each effect, NAI and TAI and a Decision Support Overlay Matrix (DSOM). The commander then makes a decision of how each effect, NAI and TAI is to be resourced, which leads to the production of the final DSOM. Again, the SA requirements during Question 5 include the actions and effects specified by the commander and the outputs of Questions 1, 2 and 3 and 4 and also information related to the capability (e.g. combat effectiveness) and available resources (and state of the available resources) that the friendly forces has.

Question 6 focuses on the time and location of each course of action, i.e. when and where do the actions take place in relation to one another? To determine this, a synchronisation matrix is produced, which includes a statement of the overall mission and the concept of operations and then a breakdown of events related to time, including enemy actions, and friendly force components activities and decision points. Key SA requirements here include the actions and effects required, the draft courses of action derived from Question 4 and 5, likely enemy actions, locations and time and how these all relate to one another.

Finally, Question 7 involves identifying any control measures that are required for the courses of action specified. Control measures are the means by which activities are coordinated and controlled. Control measures include, amongst other things, phase lines, boundaries, fire support coordination measures and lines, assembly areas and rules of engagement. Again the key SA requirements here are the friendly force courses of action and the resultant control measures required.

At a high level, the SA requirements throughout the seven questions planning process are presented in Figure 7.1. Once the plan is completed, wargaming is used to

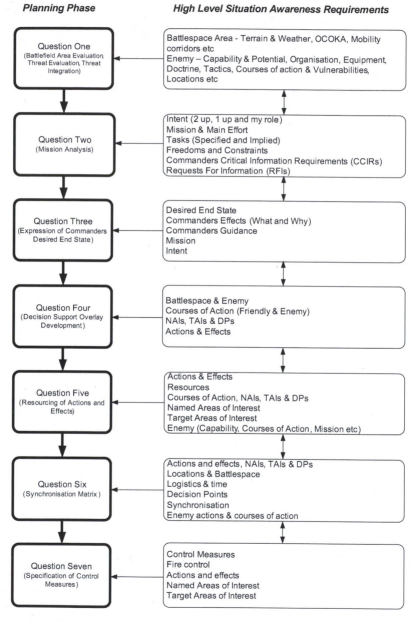

Figure 7.1   High-level seven questions planning process SA requirements

evaluate the various courses of actions specified in the plan. These are validated against the enemy's courses of action. Key decision points are also confirmed and/or refined and the coordination of assets is tested. The wargame requires members of the planning staff acting in the roles of enemy forces, friendly forces, a recorder (of information) and a referee. Several pieces of planning material are used; the decision support overlay, decision support matrix and the products of the battlefield area evaluation and intelligence preparation of the battlefield phases.

A task model of the CE planning process, depicting the main activities undertaken, is presented in Figure 7.2. Following successful wargaming, the plan is executed, outcomes are monitored and the plan is modified accordingly. Plan enactment involves responding in prescribed ways to orders received from higher command formations as they relate to information derived from the intelligence preparation of the battlefield. The command staff then direct the various force elements to engage the enemy. This is undertaken with voice/radio communications with the planning staff constantly

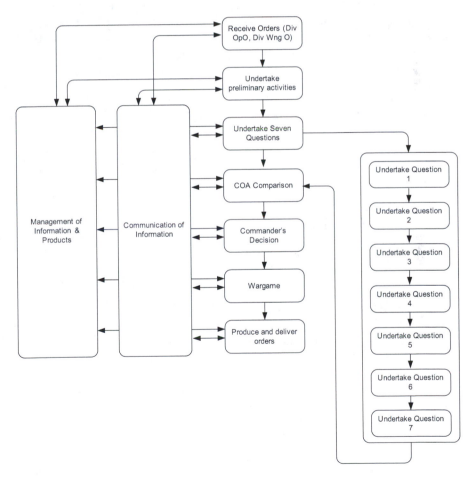

Figure 7.2   Combat estimate task model

updating dynamic aspects of the battlespace maps, as well as monitoring and where necessary cycling through the Combat Estimate technique to modify the plan.

The Study

As part of a wider analysis of the new digital system (Salmon et al., 2007) two separate training exercises used to trial the system were observed. These exercises took place at the Land Warfare Centre in Warminster, Wiltshire, UK, the first between 31 July 2006 and 4 August (hereafter referred to as 'exercise one') and the second between 25 and 29 September 2006 (hereafter referred to as 'exercise two'). The findings derived from these studies were then compared with those derived from an analysis of the 'old' paper map seven questions planning process which was undertaken previously at the Land Warfare Centre.

The observations team consisted of four researchers from Brunel University. At any one time there was a minimum of two observers logging the events of the exercise. Video and audio recorders were also used to record planning and execution activities and their associated voice communications. System logs detailing the data transmissions between the digital systems terminals were also collected post-exercise.

Methodology

Participants

Eleven participants aged between 30 and 45 years old were involved in the trial. The participant roles included an SO2 infantry staff officer, an SO2 aviation staff officer, an SO2 C3S staff officer, an SO2 Armour staff officer, an Artillery staff officer, an Engineer staff officer, an SO1 CAST staff officer, an SO2 manoeuvre staff officer, an SO1 manoeuvre staff officer and an SO2 development staff officer.

Materials

The exercises took place in a large single room partially separated with six foot high partitioning (see blue sections in Figure 7.3). The materials used included the digital mission support system terminals (a total of 28 were used), a projector and projector screen (for projecting the operational picture) and a smart board. In order to collect the data the analysts used video and audio recording equipment, CDM questionnaires and pen and paper. Laptop computers and Microsoft Word were used to transcribe the data collected (e.g. CDM responses, observational transcripts etc.).

Procedure

Both exercises consisted of training phases (whereby users were trained in the use and functionality of the digital system), planning phases (whereby users would undertake the

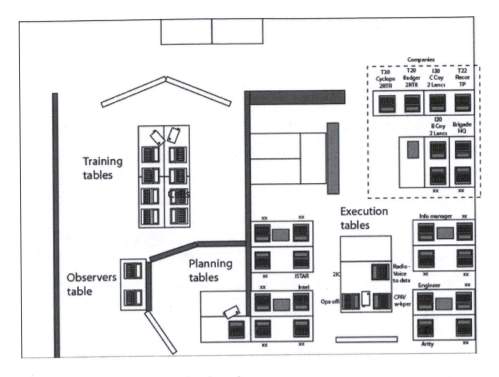

Figure 7.3   Layout of exercise domain

seven questions planning process using the digital system) and execution phases (whereby users would undertake battlefield execution scenarios using products from the planning phase). For the purpose of this chapter, only the planning scenarios analysed are reported.

During both exercises, planning operations were conducted at the planning and execution tables (see Figure 8.3). The analyst team was situated adjacent to planning tables and had access to all activities undertaken, as well as access to the trainees post exercise. Analysts observed the planning activities undertaken and recorded the voice communications between participants. Video cameras were used to record the activities undertaken. Once each planning phase (e.g. each question) was completed, the analysts conducted CDM interviews with the key agents involved.

Results

Distributed Situation Awareness

For the purposes of this research, only the results related to DSA are presented. The full analysis, including DSA, social network, teamwork and usability analyses are presented in Salmon et al. (2007).

Propositional networks were developed for the planning process activities undertaken during exercises one and two. Due to size restrictions, only the propositional networks derived from exercise two are presented. The exercise two propositional networks were developed using CDM interview transcripts. These were then compared against a HTA of the digital system seven questions process for validation purposes. The overall propositional network for the exercise two seven questions planning process is presented in Figure 7.4. Figure 7.4 also depicts the overall usage of the information elements throughout the entire seven questions process. The propositional networks for Questions 1 through to 7 follow in Figures 7.5–7.11. Within the propositional network, usage of information elements (i.e. the information required for that portion of the Combat Estimate) is represented via shading of the nodes.

The key information elements related to each phase of the planning process undertaken during both exercises were extracted by summing the links between the information elements within the propositional networks. Following Stanton et al. (2006), those information elements with six or more links were classified as key information elements. The usage of the key information elements during exercise two is presented in Table 7.1 (see page 133).

Table 7.1 presents an analysis of core information elements for each phase in the scenario; in other words, what are the essential pieces of information relating to each phase in the scenario. A total of 42 key information elements were identified from the exercise two propositional networks. Similar to those key information elements identified during exercise one, the activation of these key objects changes in type and structure. For example, information elements such as terrain analysis, locations, overlays and movements predominate in the early phase of the process, whereas resources, formations, equipment and capabilities dominate later phases. A total of nine elements were found to be used during every phase of the planning process. These included the mission, brief, commander, effects, enemy, and time. These elements are presented, along with the causal links between them, in Figure 7.12.

The information elements presented in Figure 7.12 represent those elements that were used during each planning phase; the elements in question are passed from one planning phase to another and are modified based on the meaning placed on them and their combination with other information elements. For example, the information element 'effects' is likely to be used very differently and also in conjunction with different information elements during each planning phase. During Question 3, for example, the 'effects' are considered in terms of what they are and how they relate to the desired end state, whereas during Question 4 they are considered in terms of the actions required to achieve them. During Question 5 the effects are considered in terms of the resources required to achieve them and during Question 6 they are considered in terms of where and when they will occur in relation to one another. Thus the information element 'effects' is being viewed differently and used in combination with different information elements during each planning phase.

The total information element usage during each of the seven questions phases (1–7) was also calculated. This is presented in Figure 7.13.

This analysis indicates that Questions 1, 4 and 6 incur the most loading in terms of the amount of information elements that are required.

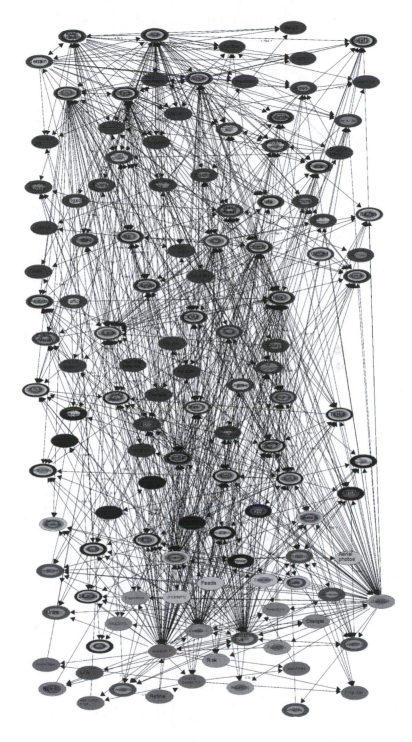

Figure 7.4   Exercise two overall propositional network depicting information element usage during each phase of the planning process

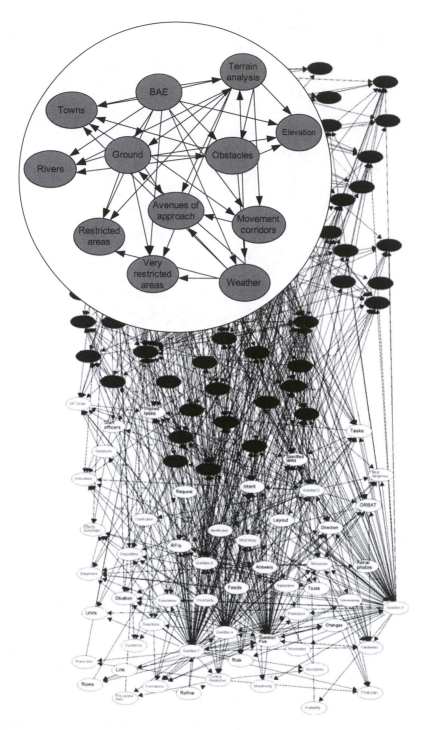

Figure 7.5   Question 1 information element usage

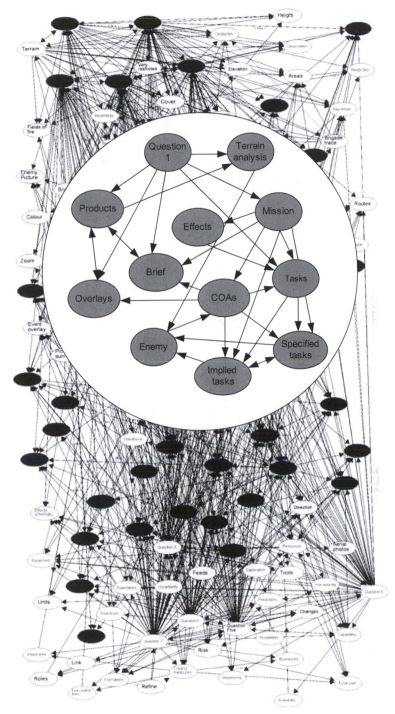

Figure 7.6   Question 2 information element usage

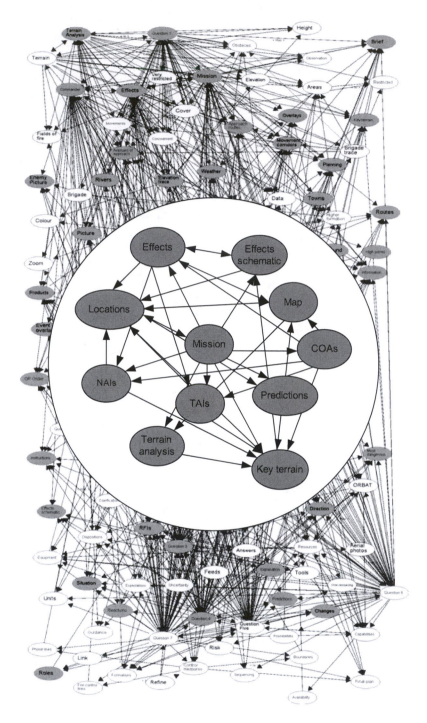

Figure 7.7   Question 3 information element usage

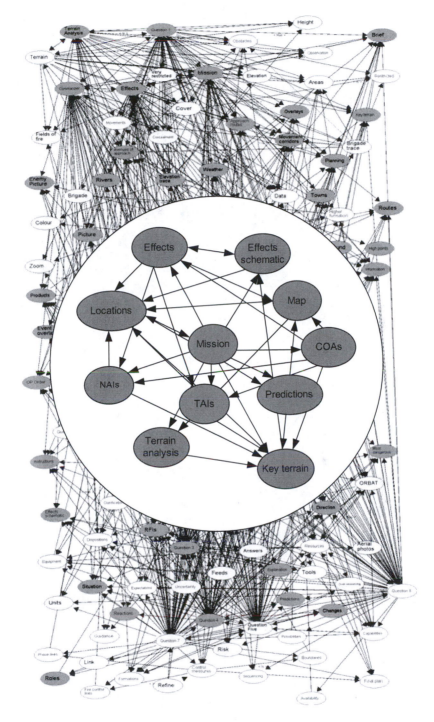

Figure 7.8   Question 4 information element usage

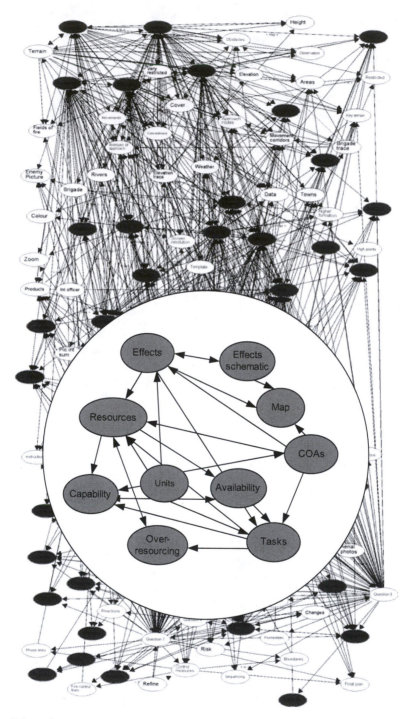

Figure 7.9   Question 5 information element usage

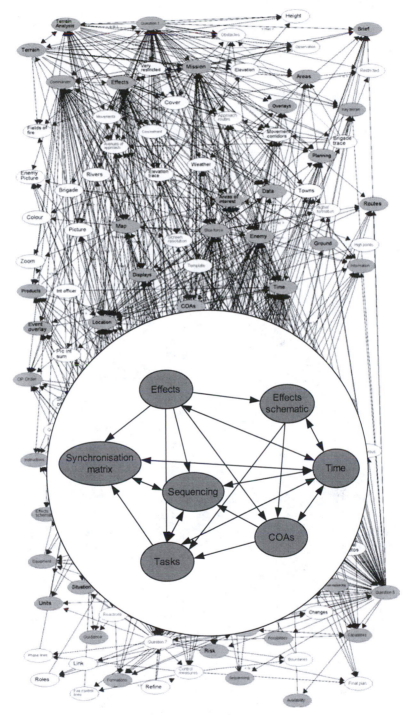

Figure 7.10   Question 6 information element usage

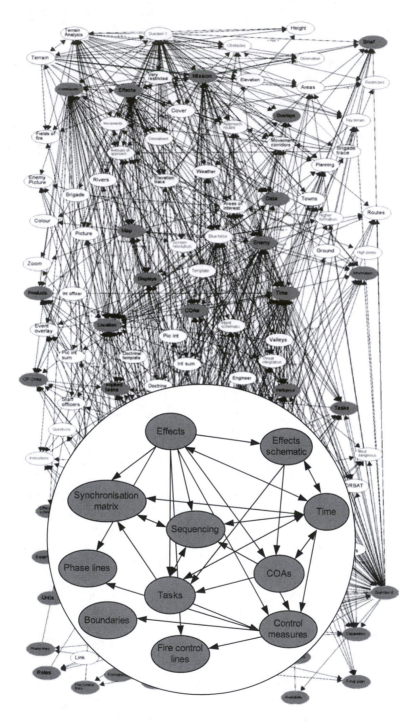

Figure 7.11   Question 7 information element usage

Table 7.1   Key information elements for exercise two seven questions scenario

Knowledge Element	Question 1	Question 2	Question 3	Question 4	Question 5	Question 6	Question 7
Terrain							
Obstacles							
Mission							
Brief							
Commander							
Effects							
Battlefield Areas							
Fields of Fire							
Movements							
Concealments							
Overlays							
Weather							
Enemy Picture							
Areas of Interest							
Routes							
Information							
Enemy							
Time							
COAs							
Blue Force							
Displays							
Locations							
Op Order							
Products							
Implied/Specified Tasks							
Intelligence							
Plan							
ComBAT							
Tactics							
Instructions							
Intent							
Effects Schematic							
ORBAT							
Direction							
Resources							
Formations							
Capabilities							
Control Measures							
Dispositions							
Equipment							
Situation							

Comparison of the Old and New Planning Processes: The Digital System versus Paper Maps

The present analysis findings were compared with a recent analysis of the analogue (i.e. paper map) seven questions planning process (see Walker et al., 2006b). Briefly, the traditional 'paper map' analogue planning process involves the conducting the seven questions using paper maps, whiteboards, flipcharts and acetate overlays. Key elements related to the plan are drawn on acetate overlays (e.g. terrain analysis, commander's effects, situation overlays etc.) and products are produced on paper, whiteboards (e.g. mission analysis) or on acetates (e.g. overlays such as the commander's effects

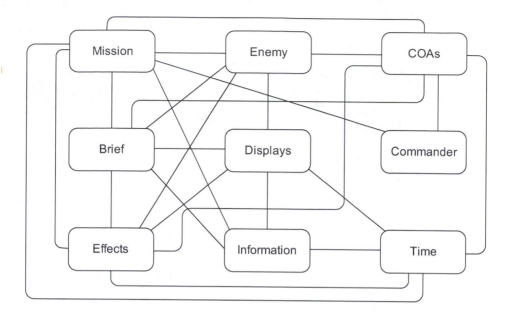

Figure 7.12 Transactive seven questions knowledge elements

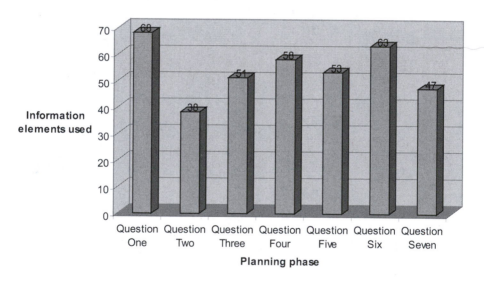

Figure 7.13 Information element usage per planning phase

schematic). The purpose of this comparison was to identify the affects that the new digital planning system had on DSA during the mission planning process. This could potentially highlight changes in the seven questions planning process which are brought about by the digital system, i.e. instances of technology affecting process and also changes in DSA requirements brought about by the digital system, i.e. increased or decreased DSA requirements. For example, it may be that the seven questions process is lengthened (i.e. more task steps) or made more difficult (increased workload) by the digital system. Conversely, it may be that the process is significantly reduced or made easier by the digital system. In relation to DSA, it may be that the introduction of a software support system may increase DSA requirements (i.e. the information that planners use and need to know about) or hinder DSA through usability and system performance problems. A brief summary of the comparison of the two analyses, labelled hereafter as CAST (the traditional process) and the digital system (the new process using the newly developed software) is presented below.

The propositional networks taken from exercise one and exercise two analyses were compared to the propositional networks from the CAST analysis. This comparison allows the identification of differences between the two processes in terms of the types of information required and the amount of information used during each of the seven question phases.

The total information element usage during each planning phase (Questions 1–7) for the CAST planning process and the digital system exercises one and two was also calculated. This is presented in Figure 7.14.

The comparison indicates that for the seven questions planning process, the digital system seems to increase DSA requirements and thus load planners more in terms of the information elements required than the CAST process does. That is, more information elements are required overall when using the digital system to undertake the seven questions. This means that the individuals, sub-teams and system 'need to know' about more pieces of information when undertaking the planning process using the digital system. Taking a closer look at the content of the propositional networks, the increased information element usage found during the digital planning processes appears to be a

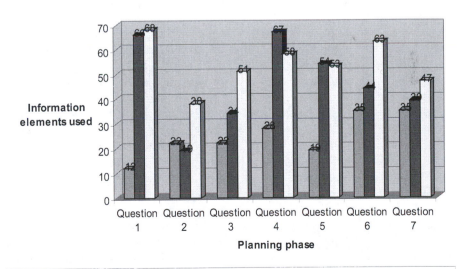

Figure 7.14 Information element usage comparison

function of three main factors. Firstly, the different features and tools within the digital system seem to add information elements to the process, i.e. users need to know more as there are additional functions and tools contained within the digital system. This means that users are having to think about the system and its features in addition to the things that they think about in order to undertake the planning process. For example, features related to the components of the digital system such as user defined overlays, drawing tools and overlay viewers are all present in the propositional networks derived from the digital planning process but are not used during the CAST manual process (since the tools in this process consist of pens, paper maps, stickies and acetates).

Secondly, various usability issues that the users had with the digital system served to create further information elements, or things for the users to think about when using the system. This was demonstrated in the exercise two analysis when the information elements 'screen', 'resolution', 'zoom' and 'display' were present in the propositional networks, which suggests that problems with the screen resolution of the digital system may be affecting the planning process since the users have to consider which resolution to use when undertaking planning tasks. These additional digital system oriented information elements are not present within the CAST manual process propositional networks (since the tools used to support this consist of paper maps, pens and stickies).

Thirdly and finally, the difference in the number of information elements used between the two processes could also in part be due to the more in-depth data (e.g. SOIs, CDM interviews and verbal transcripts) that was used in the construction of the digital system propositional networks. This may mean that the propositional networks developed for the digital system planning processes are more detailed and thus contain more information elements.

It was concluded that the main advantages (in relation to DSA) of using the digital system over the traditional paper map process include that the digital system supports rapid communication of large quantities of information over potentially very large networks and distances. In the paper map system physical products have to be delivered between units, whereas the digital system uses a digital messaging capability to deliver products. Communications using the digital system are thus quicker and one communication can simultaneously be transmitted to many. The digital system also does not require receiving agents to answer, which is advantageous, and it is highly auditable since it keeps system logs of communications. The advantages in communications have key implications for DSA during operations, since potentially DSA-related information can be communicated more quickly and to more people, which in turn should increase the tempo of DSA development.

There were, however, some key disadvantages associated with the digital system that require further investigation. One of the most interesting issues related to this research is that the digital system does not, in its present format at least, consider the variety of different roles and SA requirements that are present within the land warfare planning system. Instead, the system has a common interface for all users and contains the same tools and functions regardless of role, goals and the tasks being performed. This means that users are often presented with information, displays, tools and functions that they do not require and are unlikely to use, which in turn could potentially affect planning performance and tempo. Individuals with very specific roles and SA requirements therefore have to locate and use only the parts of the tool that they require. Whilst this aspect of the system was not investigated in this study, it is the focus of the study presented in the following chapter.

Also of importance is the fact that, in its present form, the communications network remains largely unstable and the system often breaks down under a high communications load. The system also requires extensive training for users to become efficient in its usage and the error potential of the system is extremely high. In terms of the presentation of information by the digital system to its users, there are significant problems with screen resolution, legibility, viewing area size and icon size. These issues with the digital system are investigated further with regard to their impact on DSA in the following chapter.

Discussion

The purpose of the study described in this chapter was to analyse DSA during the seven questions mission planning process and to analyse the impact of the newly developed digital software tool on DSA during this process. Within the wider context of this overall research, the intention was to explore the impacts on SA that technological systems and digitisation are likely to have during military planning and execution activities.

Propositional networks were developed from exercises one and two for the digital system supported seven questions planning process and were compared to propositional networks developed for the traditional CAST paper map planning process. The results indicate that planners used more information during the digital system supported planning process than they did during the traditional paper map-based planning process. From this it is tentatively concluded that, when using the digital system, users are required to think about or know more in order to undertake the planning process successfully. Interestingly the increased information elements were not a function of the enhanced capability for acquiring and communicating task-related information around the system (one of the proposed tenets of NEC systems), rather they appeared to be a function of the additional functionality offered by the tool and also various usability and interface design problems extant within the digital system (i.e. problems with mapping, screen resolution, zoom and the digital system interface). To demonstrate, the propositional networks derived from exercise two contain the information elements 'display', 'screen' and 'resolution', which indicates that the users are having to think about features such as the way in which planning information should be displayed and the screen resolution used when undertaking the planning process. This is something that did not come out of the analogue CAST analysis. Looking at the data further, the following response was given in the CDM interview for Question 1:

> I suppose we are used to dealing with maps but the problem with the screen resolution is that if you zoom in to look at the ground in detail you've got … the resolution isn't great so you lose the actual picture of the ground because its all lines and contour lines

but if you zoom out enough so that you can see enough ground to get the shape of the ground you lose the detail so you can't see specific areas.

This again indicates that users had to think about the technology as well as the task that they were performing. This potentially not only adds to user workload but also could take their attention away from the task in hand. This finding alone has interesting implications for the digitisation of military processes, since it seems to suggest that inappropriately designed digital mission support systems may increase the information that users need in order to develop and maintain appropriate levels of SA, which in turn may reduce tempo and increase workload. Also noteworthy is that in this case these increases in information requirements were not task related; rather they were related specifically to the digital system. Thus, it may be that inappropriately designed digital mission support systems could hinder rather than aid DSA during the planning process. It also suggests that system DSA may become denser in terms of the information comprising it and so individuals, teams and systems effectively need to know more in order to achieve levels of SA appropriate for efficient task performance. It is worth noting here that more information does not necessarily mean 'more SA' and further investigation into this aspect of digitisation is recommended.

The identification of key information elements also indicated that issues associated with the digital system affected DSA during the planning activities analysed. It was concluded from the 'key' information elements analysis that the critical information elements related to DSA during the seven questions planning process included features related to the battlefield area (e.g. battlefield area evaluation, terrain analysis, locations), the mission (e.g. effects, mission analysis, commander's direction), the enemy (e.g. enemy, threat evaluation, courses of action), and execution activities (e.g. targets and areas of interest, time and overlays). Further, a series of 'transactive' information elements (i.e. used during every phase of the planning process) was identified. These included the battlefield area, the commander, the commander's direction, overlays, the plan, time, displays, effects, information, the mission, the enemy, courses of action and the brief. Interestingly, the presence of 'displays' within the key information elements was a result of the users having various problems with the display of information by the digital system. For example, the screen resolution and display size often led to the users not being able to view the entire battlefield area and losing context on the battlefield when zooming in and out of the battlefield area.

In terms of the different phases of the planning process, the propositional networks indicate that during exercise one, Questions 1 and 4 required the most information elements (i.e. the agents needed to know the most things) to undertake (66 and 67 information elements respectively), followed by Question 5 (55), Question 6 (44), Question 7 (39), Question 3 (34) and Question 2 (19). During exercise two, Questions 1 and 6 (68 and 63 information elements respectively) required the most information, followed by Questions 4 and 5 (58 and 53), Question 3 (51), Question 7 (47) and Question 2 (38).

In comparing the two planning processes, the analysis indicated that the SA requirements remain the same whether the process is undertaken manually using paper maps or digitally using the digital system. Albeit at a high level, planners essentially need to understand the key features of the battlespace area, enemy and the overall friendly force mission and commander's effects, use this information to identify likely enemy courses of action and then develop, resource and synchronise their own courses of action. The information required to support planner SA requirements is therefore derived from maps of the area in question, intelligence of the enemy, mission orders, historical information and planner experience, planning products from other questions. The main difference between the two systems therefore relates to what information is presented to the users and how it is presented, and also how the planning products are constructed using both systems.

In the paper map system, maps of various scales and acetate overlays are used to present battlefield area-related information and develop planning products. For example, the commander's effects schematic is manually drawn on acetate and placed on the appropriate area on the paper map. The effects schematic is then presented to the planners on a bird table. Within the digital system, the user has to draw the effects on a user-defined overlay using the digital system's drawing tools and the effects schematic product is presented on a smart board display. The two main differences here relate to the digital system tools and the resolution of the maps on the digital system. Firstly, the digital system's drawing tools process is a convoluted one and so tempo is lost when using the digital system to draw and construct overlays such as the effects schematic. Secondly, the resolution of the maps used on the digital system is problematic and users often cannot see the ground in enough detail, and when they zoom out, they lose context of where they are in the battlespace.

The SA requirements for the planning process are also interesting in that it appears that some of the information required is not presented, rather data is presented that requires the planners to make assumptions (based on experience and other information) in order to develop detailed SA. For example, during the threat evaluation phase the planners have to identify, based on the terrain, enemy doctrine and past activities, probable enemy courses of action. This requires that all the information from the battlefield area evaluation be assimilated along with historical information and experience in order to make assumptions on the enemy's likely courses of action. Thus portions of the seven questions process is completed using a combination of current situational information (e.g. the terrain and the enemy), experience (e.g. appropriate courses of action, resourcing and control measures) and historical information (e.g. the enemy's previous activities and doctrine). This information is then used to develop courses of action. These features of the planning process have interesting connotations for the design of a digital support system, namely that it needs to present current situational and also historical information, support user experience and also support the creative planning process.

In terms of the SA requirements for the different planners involved in the seven questions process, there are similarities in terms of the SA requirements across positions. For example, all positions require an understanding of the overall mission and the commander's effects and desired end state and an understanding of the features of the battlespace area and the enemy being faced. However, critical differences exist between the SA requirements of each component involved in the planning process and it is apparent that the digital system does not support these different SA requirements.

Rather the system is the same in terms of the information that it presents, the format in which it presents the information and also the menus and tools available, regardless of who is using the system. It is apparent that a more role-specific design could have been adopted whereby different users could tailor the interface, tools and information presented based on their role and SA requirements. This aspect of the digital system is explored in more detail in the next chapter.

Differences between the SA requirements of each user are more pronounced in the format that the information is required in, the level of detail that the different positions require the information to be in and in the ways in which the different positions use the SA-information that is presented to them by the system.

It is clear that further investigation relating to the impact that the digital system has on DSA is required. Of particular concern is the systems effect on DSA during real world activities, as opposed to training scenarios and so further evaluation of digital system during in-the-field exercises is required.

Is it Really Better to Share? Analysis of a New Digital Mission Support System and Implications for System Design

Introduction

The previous chapter presented a DSA-based analysis of a new land warfare digital mission support system and focused on the impact on DSA that digitising the seven questions mission planning process had during land warfare mission planning training exercises. The findings suggested that the new digital system was adversely affecting the planning process and DSA during the planning activities observed. The purpose of this chapter is to further investigate the mission support system with regard to its effect on DSA during operational activities. Whilst the previous chapter focused on the effect that the new system had on DSA in comparison with the old paper map system, the analysis presented in this chapter focuses more on the way in which the system is designed and how this affects DSA acquisition and maintenance during operations.

The HFI-DTC was tasked to provide an independent evaluation of the digital system during an operational field trial that was conducted by the Command and Control Development Centre (C2DC) during November 2007. As part of this overall HF analysis, the impact of the digital system on DSA during mission planning and execution activities was assessed. The findings derived from this analysis are presented in this chapter. In addition, the aim of this chapter is also to investigate the concept of compatible SA further and demonstrate how the DSA theory and propositional network methodology can be used to evaluate system design and also to inform system design and redesign.

System Aspirations

Much has been said on the enhanced capability that the digital mission support system in question is likely to provide. One of the critical success factors stipulated prior to the operational field trial (cited by Macey, 2007) is that it enhances user SA during planning and execution activities. Further, amongst other things, SA is one of the key areas in which the proponents of the system claim it will provide specific enhancements to planning and execution activities. For example, a promotional video regarding the new system makes the following claims (with emphasis added).

> An important benefit of [the system] is *improved situational awareness*. That is answering the age old soldier's questions: Where am I? Where are my mates?

Units that convert to [the system] will have *greater situational awareness* than before, and future increments delivered at regular intervals will improve that capability. Another fundamental difference that [the system] has brought about is the ability to transmit data which introduces marked improvements in the speed and accuracy with which information can be sent processed and acted upon.

Situational awareness means we can make decisions faster in the head quarters, a critical element of that is the secure communications. Which means that if we need to confirm anything we can do so quickly but very often we don't need to at all because the information is pushed to us already through the *situational awareness* picture. Because we know where people are we can take decisions on fires and manoeuvre quicker and that makes us a much more capable brigade.

To summarise, a number of claims have been made regarding the new digital system and its impact on SA development and maintenance during mission planning and execution activities, including that, during land warfare operations, it provides:

- greater timeliness of the passage of information;
- greater accuracy in the passage of information;
- improved SA of own position;
- improved SA of friendly positions;
- improved SA of enemy positions; and
- improvements in the speed and accuracy with which information can be sent, processed, and acted upon.

Land Warfare Mission Planning and Execution and Distributed Situation Awareness

It is no surprise that the concept of SA emerged within a military context (Endsley, 1995a), where it is an integral component that can make the difference between life and death and victory and defeat in conflicts. SA is a critical commodity in the military land warfare domain, where distributed teams have to understand dynamic, information rich, uncertain, rapidly changing environments and elements and plan and execute activities against multiple adversaries working to defeat them. Rasker et al. (2000; cited in Riley et al., 2006) also point out that command and control teams need to perceive, interpret and exchange large amounts of ambiguous information in order to develop and maintain the SA required for efficient performance. Further, unlike civilian domains, military systems also have the added complexity of an adversary attempting to inhibit the development and acquisition of SA during operations. During mission planning, inadequate or erroneous SA can ultimately lead to inadequate or inappropriate plans and courses of action and during battle execution degraded SA can lead to loss of life, failed missions, and in the worst case, loss of overall conflicts. In a military context, SA has been defined as:

the ability to have accurate real-time information of friendly, enemy, neutral, and non-combatant locations; a common, relevant picture of the battlefield scaled to specific levels of interest and special needs. (TRADOC, 1994; cited in Endsley et al., 2000)

The level of SA afforded by both the command and control process (i.e. the seven questions) and the command and control system (i.e. the digital mission support system) during warfare planning and execution is therefore a critical factor in mission success. In relation to digital planning and execution tools, Endsley and Jones (1997) suggest that the way in which information is presented influences SA by determining how much information can be acquired, how accurately it can be acquired and to what degree it is compatible with SA needs (Endsley and Jones, 1997). Therefore how SA related information is assimilated, what information is presented by the system and to whom, the format in which is it presented and the timeliness of its presentation to users are pertinent issues when considering DSA and the assessment of the digital system. All of these issues ultimately relate to the overall level of DSA that is afforded by the digital system.

Mission Planning

In addition to being complex, time dependent and subject to uncertainty (Riley et al., 2006), mission planning is contingent upon planners having an accurate SA of current and future events. Riley et al. (2006) point out that SA is a critical commodity during planning and course of action development and also that the effectiveness of the military planning process is highly dependent upon an accurate assessment of the situation. A plan can be defined as a proposed sequence of actions to transform a current state into a desired state (Klein and Miller, 1999). Within command and control environments such as military land warfare, planning is characterised as dynamic, collaborative, highly time dependent and subject to high levels of uncertainty (Klein and Miller, 1999; cited in Riley et al., 2006). In the land warfare context, the information comprising SA is inherently complex and so the planning process and system used should be designed to facilitate the acquisition and maintenance of accurate, up-to-the minute SA. Discussing the complexity associated with supporting SA during mission planning, Riley et al. (2006) point out that:

> Much of the information that soldiers use to develop plans and orders is rife with uncertain data. Battlefield information is complex, and identifying what information is confirmed and what is suspected or assumed can frequently be difficult. Furthermore, modern military operations can involve vast amounts of such incongruent data coming from numerous sources (both digital sensor and human intelligence). These data must be perceived, filtered, analysed, and effectively exploited for making decisions and formulating COAs in a timely manner. (Riley et al., 2006, p. 1142)

In order to facilitate DSA then, the digital system should present accurate and appropriate information to the appropriate users in a timely manner and also in a manner which supports the efficient and timely acquisition and maintenance of DSA.

Battle Execution

Accurate, up-to-the minute DSA is required for efficient battle execution, which involves control of activities in the field by the operations cell of the brigade or BG. Within land warfare, it is repeatedly emphasised that plans do not survive first contact with the enemy and so accurate and up to the minute DSA is required in order to control and direct activities on the battlefield. Lawson's (1991) model of command and control suggests that data is extracted from the environment, processed and then compared with the desired end state. Discrepancies between the current state and the desired end state serve to drive decisions about how to move towards the desired end state once more. These decisions are then turned into actions and communicated to the forces in the field. DSA is obviously of critical importance here, since it is the accurate understanding of the current situation that is compared to the desired end state. Of primary importance then during battle execution are the provision of an accurate and up-to-date local operational picture (LOP) and also the presence of communications links for facilitating DSA acquisition and maintenance.

Digital System Distributed Situation Awareness Assessment

The aim of the analysis presented was to assess the impact on DSA that the new electronic mission support system had during the activities analysed. Based on the appropriate literature (e.g. Endsley and Jones, 1997; Riley et al., 2006; Salmon et al., 2008) it was possible to postulate a number of critical SA-related requirements that the digitised mission support system should satisfy in order enhance, rather than inhibit, SA acquisition and maintenance during mission planning and execution activities. These are summarised below:

1. *Support for the range of different user SA requirements.* The research described up to this point has demonstrated that, in complex collaborative systems, different team members have different, but compatible, SA requirements. The nature of mission planning and execution activities in the military land warfare domain is also such that different agents with different goals and roles require different types of information or use the same information very differently. Digitised mission support systems therefore need to be able to present the appropriate information, in an appropriate format, to the appropriate users at appropriate times. Endsley and Jones (1997) suggest that the key to SA and information dominance in future warfare is in getting the right information to the right person at the right time in a form that they can quickly assimilate and use. The system should therefore possess the capability to present compatible SA-related information to its users; further users should also be able to tailor the system to there own specific needs (i.e. be able to customise the interface, tools available, information presented and format of the information presented).

2. *Presentation of SA-related information in a timely manner.* The temporal nature of DSA and the subsequent importance of keeping DSA up to date was emphasised by the findings from the studies presented in Chapter 5 (energy distribution) and

Chapter 7 (military mission planning). Both studies highlighted the importance of communicating information around a system in a timely manner in order to keep DSA up to date with the real state of the world; untimely information distribution ultimately leads to diminished DSA. Planning and execution activities are time critical and operational tempo is one of the key factors in the success of land warfare missions. SA-related information should therefore be presented to users in a timely manner, without any delay, in order to enhance planning and execution tempo.

3. *Presentation of accurate SA-related information.* An efficient level of DSA is dependent on the exchange of accurate SA information between agents within the collaborative system. Likewise, efficient planning and battle execution is contingent on an accurate understanding of the situation; it goes without saying that the information presented by the mission support system should be up-to-date and accurate. Bolia et al. (2007) point out that inaccurate data can emerge from erroneous assumptions made by data fusion algorithms (e.g. a data fusion algorithm deciding that two sensor inputs represent a single entity when they in fact represent two different enemy vehicles); from deliberately fabricated data being fed into the network; or from data that is temporally no longer correct (Bolia et al., 2007). The SA-related information presented by the system should therefore be accurate and free from spurious or out of date data.

4. *User trust in the system.* The users of the system should at all times implicitly trust the SA-related information that is presented to them. Endsley and Jones (1997) suggest that confidence in data is a particular problem in combat environments since information is often dated, conflicting, interpreted incorrectly or patently false. They point out that the amount of confidence an operator has in the accuracy and completeness of information received is a critical element of SA development and maintenance (Endsley and Jones, 1997).

5. *Support for the level of DSA that is required for efficient, timely and effective mission performance.* Team performance is most effective when there is good DSA throughout the system in question (Stanton et al., 2006; Salmon et al., 2008); ultimately then, mission support systems should be judged on the overall level of DSA that they afford and whether or not they facilitate a level of DSA which enables the achievement of the mission under analysis.

In addition to assessing the nature of DSA (in relation to DSA theory) during military land warfare planning and execution activities, the purpose of this analysis was to evaluate the digital system in terms of the requirements specified above. Each requirement formed a sub-hypothesis with the positive outcome as the expected outcome.

Methodology

Design

The study involved a live observational study of an operational field trial of the digital mission support system. The three-week trial involved a fully functional Division,

Brigade and Battle Group (BG) undertaking mission planning and execution activities using the new digital system. The trial was set up specifically in order to test the new system and closely represented a real world operational situation.

Participants

The participants involved in the study were the army staff working in the Brigade and BG teams involved in the operational field trial. The Brigade and BG HQs analysed consisted of the following cells for the staff to work from; G3 Operations, G5 Plans, G6 Operations, Combat Systems Support Operations (CSSO), Air Aviation, G2 Intelligence, ISTAR, I-Hub, Artillery and Engineers. Due to the nature of the study and data restrictions, it was not possible to collect participant demographic data. A diagram depicting the Brigade and BG HQs and the component cells is presented in Figure 8.1.

Materials

The materials used for this study included the digital mission support system terminals, the resources used by the Division, Brigade and BG throughout the trial, including the materials at each HQ (e.g. maps, pens, VHF radios, acetates, tables, chairs, smart boards, standard operating instructions etc.) and also on the battlefield (e.g. vehicles, equipment, weapons etc.). The materials used by the analysts to collect the data included notepads and pens, digital cameras and audio recording equipment.

Figure 8.1   Brigade/Battle Group HQ layout showing component cells

Procedure

The procedure involved two main components: firstly, an analysis was undertaken in order to identify the DSA requirements of the Brigade and BG members during the mission planning process and, secondly, an analysis of DSA during the mission planning and execution activities observed was undertaken.

For the SA requirements analysis, a HTA was constructed for the planning process using data derived from standard operating instructions and interviews with SMEs. The HTA description was then refined over the course of the field trial on the basis of observations and further interactions with SMEs. The SA requirements of the different team members involved were then extracted from the HTA description and were used to construct propositional networks, depicting DSA requirements, for each team member.

For the analysis of DSA during mission planning and execution activities, a total of six analysts located within the Brigade and BG HQs undertook direct observation of the planning and battle execution activities over the course of the three-week trial. The analysts were located within the HQs during both planning and execution activities. The data recorded during the observations included a description of the activity (i.e. component task steps) being performed by each of the agents involved, transcripts of the communications that occurred between agents during the scenarios, the technology used to mediate communications, the artefacts used to aid task performance (e.g. tools, computers, instructions, etc.), the temporal aspects of the tasks being undertaken (e.g. time undertaken, time available and time taken to perform tasks) and any additional notes relating to the tasks being performed (e.g. why the task was being performed, what the outcomes were, errors made, impact of the system on task etc.). Analysts were also given access to planning products, SOIs, logs, briefs and subject matter experts throughout the field trials. To back up the data collected during the observations the analysts frequently held discussions with the participants and SMEs. Based on the data collected, propositional networks were developed for the mission planning and battle execution activities observed.

Results

Situation Awareness Requirements Analysis

The SA requirements of each of the different team members (Brigade and BG) involved were extracted from the HTA that was developed for the planning process. These were then used to develop propositional networks depicting the SA requirements of the different team members during each of the seven questions planning phases. To demonstrate the differences in SA requirements across the Brigade/BG we focus here on Question 1 of the mission planning process. Question 1 involves the use of maps to undertake the battlefield area evaluation, which involves the terrain analysis, threat evaluation and threat integration processes. The terrain analysis component requires an assessment of the effects of the battlespace on enemy and friendly operations and involves the identification of key terrain and likely mobility corridors, avenues of approach and manoeuvre areas; threat evaluation involves identifying the enemy's likely *modus operandi* by analysing their equipment, tactical doctrine, combat effectiveness, past operations and their strengths and weaknesses; threat integration involves combining the battlefield area evaluation and threat evaluation outputs in order to determine the enemy's intent and how they are likely to operate. This results in the identification of Named Areas of Interest (NAIs) and likely enemy courses of action. A propositional network representing the BG systems DSA requirements during question 1 of the planning process is presented in Figure 8.2.

The network depicted in Figure 8.2 represents the Bde planning system's DSA requirements during Question 1 of the mission planning process. This overall view tells us what it is that the BG planning system needs to know during question 1 of the planning process. For the concept of compatible SA, however, we have to delve deeper. Figures 8.3 to 8.8 (pp. 150–155) present the DSA requirements of each of the key BG staff members involved in undertaking Question 1 of the planning process. The extracts focus on how each cell uses and views the network of information elements underlying the systems DSA. As represented by Figures 8.3 to 8.8, each planning cell is required to have a very different 'awareness' of these information elements as defined by their goals and their role and responsibility during the Question 1 planning phase. The distinct, but compatible views on the same situation that each agent involved requires suggests that, although they are at times using the same information, they use and view that information very differently.

For example, during Question 1 the engineer cell is primarily concerned with developing SA of the ground, friendly and enemy forces use of the ground and the impact that the ground is likely to have on friendly and enemy operations. The intelligence cell, on the other hand, is primarily concerned with understanding the ground and what this means with regard to the threat posed by the enemy, including the enemy's capability, strengths and weaknesses, doctrine and subsequent modus operandi. The Combat Service Support Officer (CSSO) is involved in maintaining levels of combat effectiveness, and so in relation to the ground and enemy and friendly force operations needs to consider how casualties and damaged equipment will be removed from the battlefield and treated, repaired and replenished. Important considerations in relation to the ground therefore include routes in and out of the battlefield, potential bottlenecks and medical sites and also what the repair and replenishment requirements are likely to be. The G6 (comms cell) focus mainly on the ground and its impact on friendly and enemy communications. They therefore need to consider communications assets, requests for communications capability, the actual communications capability of the enemy and friendly forces and any constraints imposed on communications by the ground. The air/aviation cell is primarily concerned with the ground and its impact on the air defence and air attack activities of the friendly and enemy forces. Finally, the chief of Staff (COS) requires an appreciation of the overall Question 1 battlefield area evaluation and what the enemies most likely and most dangerous courses of action are likely to be with the ground in question. The COS also focuses on friendly force in terms of opportunity and response to the enemies COA, along with the impact of the ground and enemy on the friendly force modus operandi.

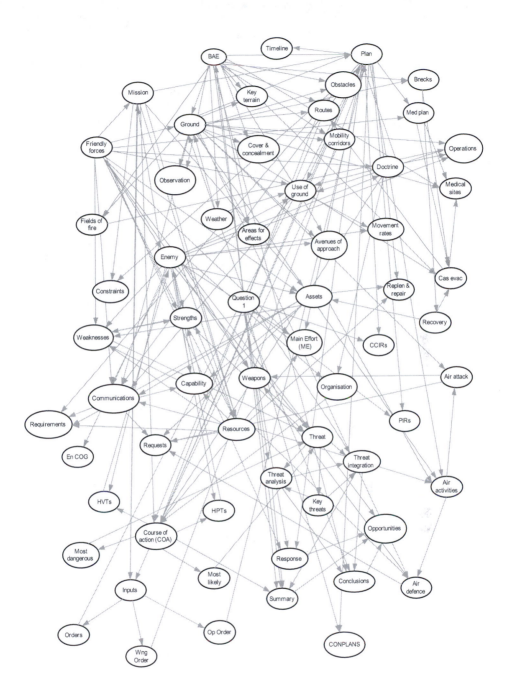

Figure 8.2 Question 1 propositional network

Note: The figure represents Bde planning teams distributed situation awareness requirements during Question 1 of the mission planning process.

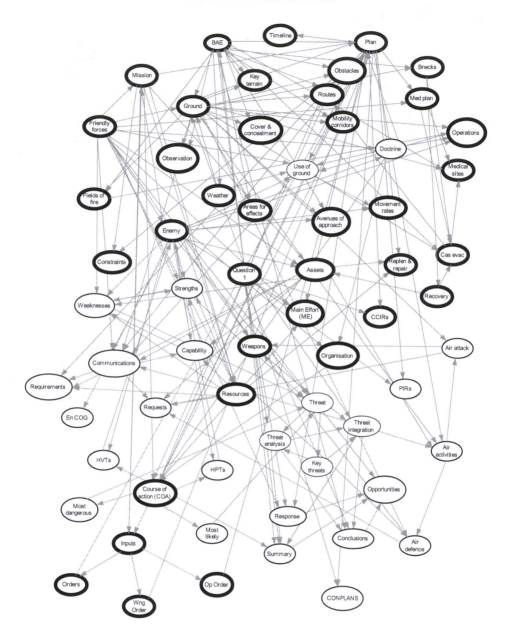

Figure 8.3   Combat Service Support Officer view

Note: The nodes circled in bold represent the Combat Service Support Officer's SA.

The examples presented in Figures 8.3 to 8.8 demonstrate how, even when different team members have access to exactly the same information (in this case information regarding the ground, the enemy and the friendly force), as a function of their specific role within the team and thus their unique goals, they use and view

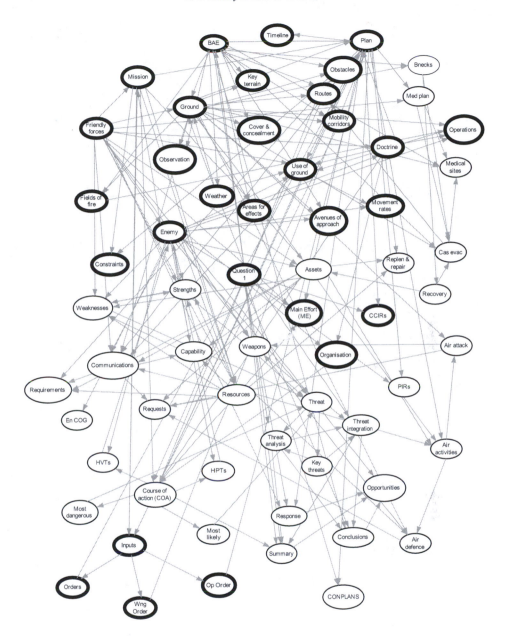

Figure 8.4   Engineer view

the information in a very different manner; it is the relationship between concepts, born out of their role in the collaborative system, that makes up their distinct SA. There awareness is, however, compatible since it connects together in order to allow production of the Question 1 products; each cell's awareness of the battlefield, the enemy and the level of threat conjugates together to form the Question 1 output. It is

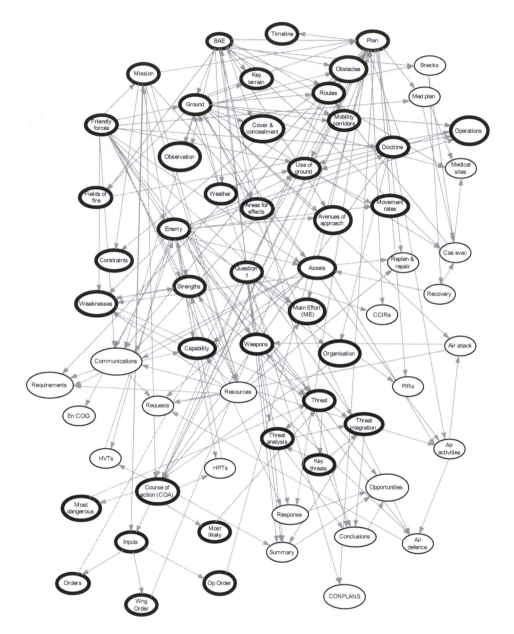

Figure 8.5   Intelligence Officer view

this unique combination of information elements and the relationships between them that make each team member's SA compatible and not shared; the very fact that an actor has received information, acted on it, combined it with other information and then passed it onto other actors means that its interpretation changes per team member.

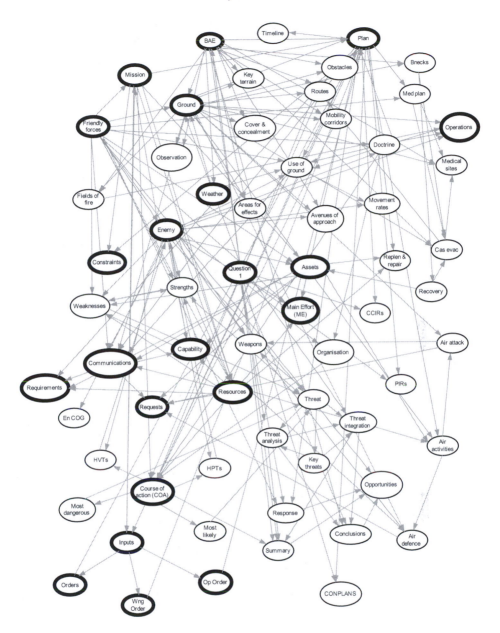

Figure 8.6   G6 (Communications) Officer view

Seven Questions Planning Analysis

The analysis of planning activities focused specifically on the seven questions planning process (see previous chapter for description). A task model of the seven questions planning process observed is presented in Figure 8.9 (see page 156).

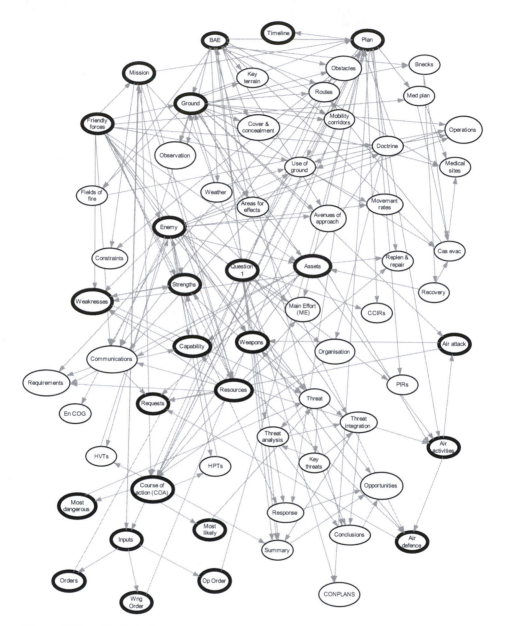

Figure 8.7   Air/Aviation Officer view

Propositional networks were developed for each question of the Brigade CE planning process observed during the operational field trial. The seven questions propositional networks for mission 1 are presented in Figure 8.10 to Figure 8.16 (pp. 157–162)

The propositional networks presented in Figures 8.10 to 8.16 depict the planning systems awareness during the seven questions planning process. It is notable that, in its

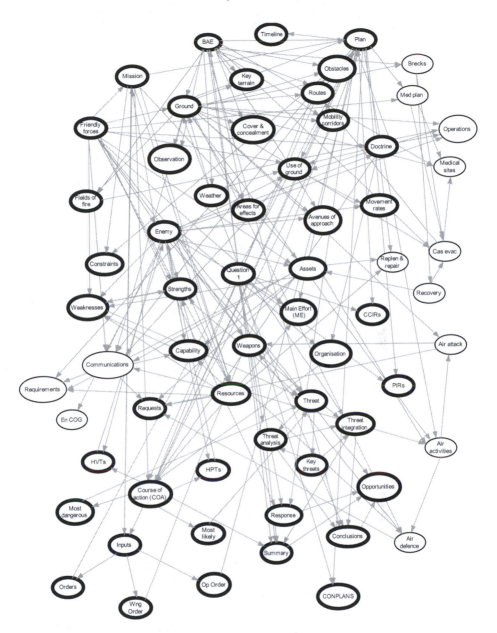

Figure 8.8   Chief of Staff view

Note: The nodes circled in bold represent the Chief of Staff's SA.

present form, the digital system presents the information necessary to support the seven questions planning process and it also provides users with the necessary functionality (i.e. tools) required to undertake the seven questions. Despite this, issues associated with the timeliness and accuracy of the information presented, presentation of the appropriate

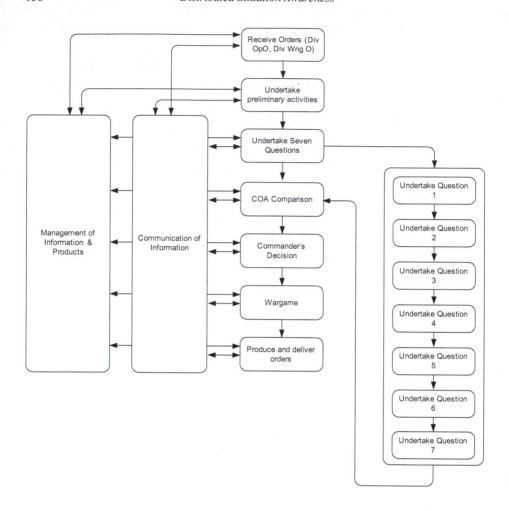

Figure 8.9   Combat estimate task model

information to the appropriate users and the usability of the digital system's planning tools all adversely affected the level of DSA that the BG held during the planning activities observed. For example, the tempo of planning was reduced due to problems with the timeliness of the information being presented. In addition, the users often had doubts regarding the accuracy of the information presented to them by the digital system, which led to them querying the information and undertaking further processes required to clarify the data, which had the effect of reducing operational tempo during mission planning. The usability issues associated with the system's planning tools (e.g. map displays, drawing tools, user defined overlays, synch matrix, maps, TASKORG etc.) also had an impact on the tempo of planning activities, since the users had various problems using them and so took too long to develop planning products. Finally, the lack of support for the different SA requirements of the different cells involved in the planning process (e.g. G2, G6, Artillery, Engineer etc.) meant that users had to find and

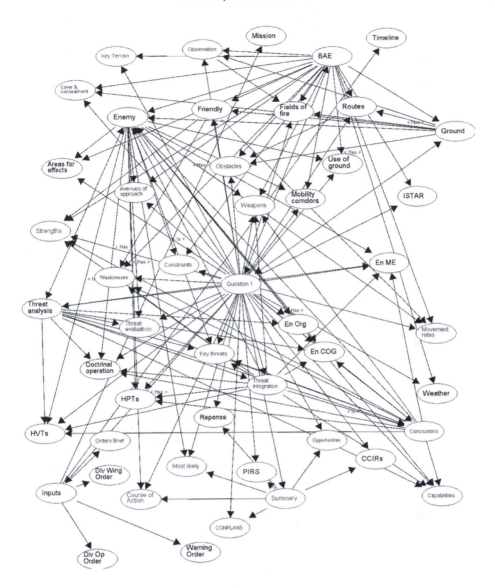

Figure 8.10 Question 1 propositional network

locate the information that they required, which was often time consuming and error prone and served to delay the acquisition of SA. These issues are discussed in detail in the discussion section of this chapter.

The key information elements were extracted from the planning execution propositional networks using the five or more links rule. For the purpose of this analysis, salience is defined as those information elements that serve as a central hub to other information elements (i.e. have five or more links to other information elements). The key information elements are presented in Table 8.1 (see page 161).

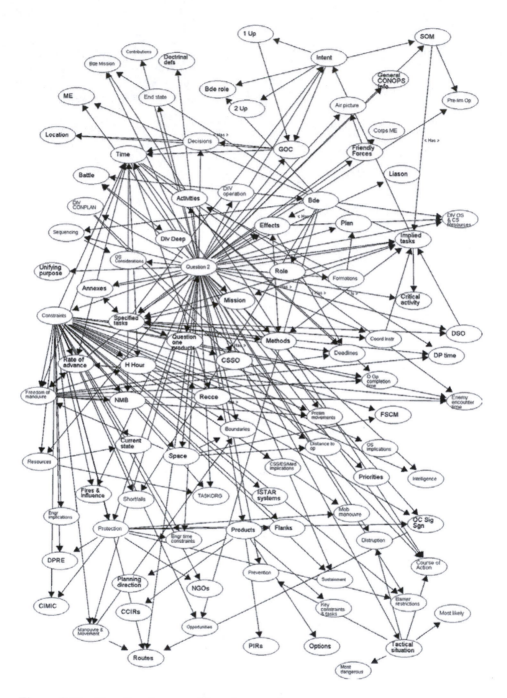

Figure 8.11   Question 2 propositional network

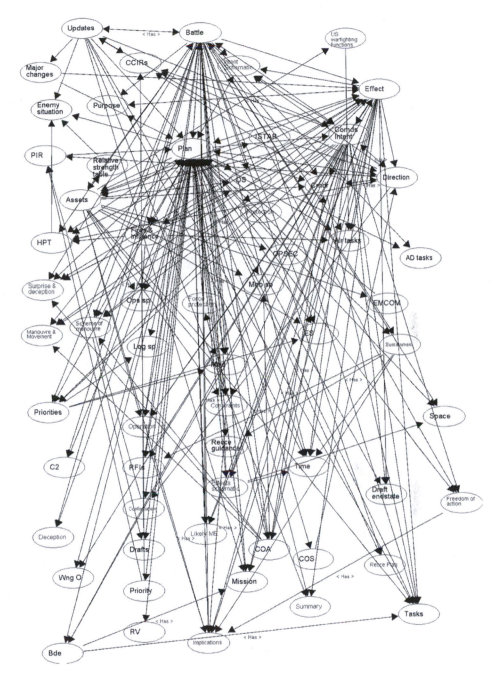

Figure 8.12 Question 3 propositional network

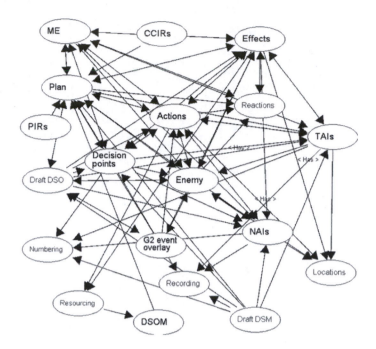

Figure 8.13   Question 4 propositional network

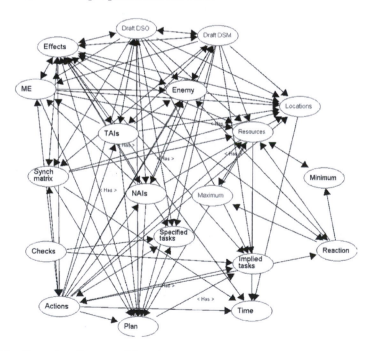

Figure 8.14   Question 5 propositional network

Table 8.1   Planning key information elements

	Question One		Question Two		Question Three		Question Four		Question Five		Question Six		Question Seven	
KEY INFORMATION ELEMENTS	BAE	Enemy	Intent	Implied tasks	Mission	Implications	Effects	ME	Draft DSO	Effects	Draft DSO	Effects	Control Measures	Effects
	Ground	Strengths	Specified tasks	Question 2	Cmdrs Intent	Effects	Plan	TAIs	Draft DSM	TAIs	Draft DSM	TAIs	Draft DSO	TAIs
	Weaknesses	Threat analysis	Time	Effects	Plan	Direction	NAIs	Locations	NAIs	Locations	NAIs	Locations	Draft DSM	Locations
	Threat integration	Question 1	Mission	Constraints	Battle	Updates	Actions	DPs	Actions	Resources	Actions	Resources	NAIs	Resources
	Weapons	Key threats	FOM	Rates of advance	CCIRs	Assets	Draft DSO	Enemy	Draft DSO	ME	Draft DSO	ME	Actions	ME
	Enemy COAs	Capabilities	Space	Protection	Fires & Influence	Cmdr			Synch Matrix	Reactions	Synch Matrix	COAs	Draft DSO	COAs
	HVTs	Inputs	Tactical situation	Products	OS	Main tasks			Implied tasks	Specified tasks	Implied tasks	Specified tasks	Synch Matrix	Specified tasks
	Summary	Mobility Corridors	Effects	Deadlines	SOM	Time			Time	Plan	Time	Plan	Implied tasks	Plan
	Conclusions	Enemy Org			Tasks	Implications			Enemy		Enemy	SOM	Time	SOM
	Enemy COG				Intent schematic	HPTs							Enemy	
					Priorities	Surprise & Deception								
					Sustainment	COA								

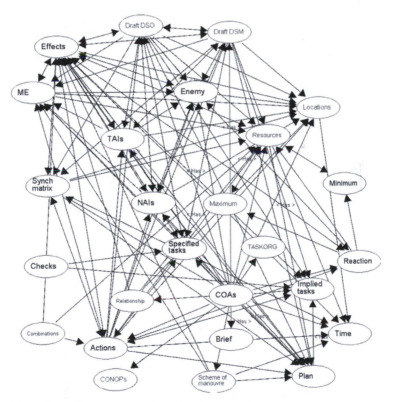

Figure 8.15. Question 6 propositional network

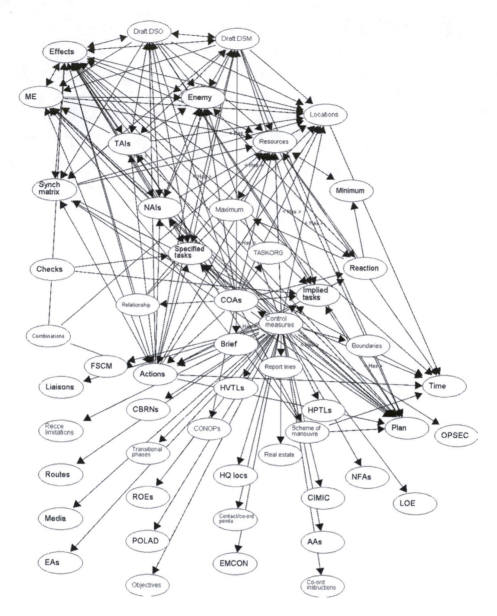

Figure 8.16   Question 7 propositional network

Battle Execution Analysis

A DSA-based analysis of the battle execution activities undertook by the Brigade during battle two (battle two occurred on Friday 16 November) is presented. Propositional networks were constructed for the battle execution activities based on content analyses of the verbal communications taking place between the key agents located at the ops table in the BG. The battle was run by the Chief of Staff (COS) and his colleagues around the operations (ops) table. The ops table layout is presented in Figure 8.17.

A task model was constructed for the battle execution activities observed. Task models provide a high-level representation of the key tasks involved. The task model for the battle is presented in Figure 8.18.

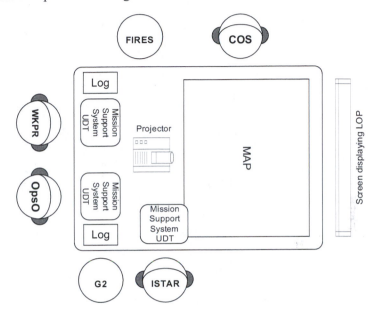

Figure 8.17   Ops table layout

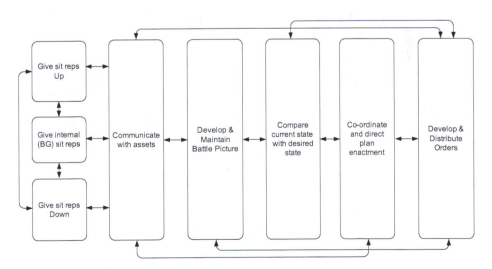

Figure 8.18   Battle execution task model

The task model shows, albeit at a high level, the critical activities that the BG were engaged in during the battle. These included developing and maintaining an accurate battle picture, comparing the current battle state with the desired end state and how the situation should be according to the plan being enacted, developing and distributing orders and coordinating and directing the enactment of the plan and communicating with assets and other elements of the command chain (via situation reports etc.). It is noteworthy that each of these activities is interlinked and is ultimately reliant on the system providing the BG with highly accurate and timely information that is required for DSA. For example, without accurate DSA of what is happening on the battlefield the BG cannot develop and maintain an accurate battle picture and thus cannot develop appropriate orders or direct the plan enactment. Also, accurate DSA of the current situation and also of the plan being enacted allows them to be compared meaningfully. Finally, the provision of accurate situation reports up, down and within the BG is also dependent upon them having an accurate picture.

A total of six propositional networks were constructed, one for the activities occurring during enactment of each of the phase lines involved in the friendly force battle plan. The phase lines represent different phases of the plan and the associated areas on the battlefield. In this case the phase lines were named after characters from the Harry Potter novels (e.g. Harry and Scabbers, Voldemort, Hagrid, Dobby, Dumbledore and Hedwig). The propositional networks are presented in Figures 8.13 to 8.18.

The key information elements were extracted from the battle execution propositional networks using the five or more links rule. The key information elements are presented in Table 8.2.

Table 8.2   Battle execution key information elements

KEY INFORMATION ELEMENTS	Harry & Scabbers		Voldemort		Hagrid		Dobby		Dumbledore		Hedwig	
	Contacts	Picture	Battle	Picture	Friendly Forces	Picture	Contacts	Picture	Enemy	Picture	Enemy	Picture
	Friendly Forces	Enemy	Friendly Forces	Enemy	Plan	Enemy	Friendly Forces	Enemy	Plan	Battle	Plan	Battle
	Plan	Situation	Locations	Situation	Voldemort	Situation	Plan	Situation	Voldemort	Assets	Phase lines	Assets
	Map	Locations	Opening	Time	Assets	Locations	Battle	Assets	Phase lines	Situation	Friendly Forces	Situation
	Op Order	Icons	Phase lines	Javelin	Phase lines	Battle	Time	Phase lines	Time	Friendly Forces	Time	Cancellation
	Obstacles	TAIs			Cancellation	Time	Marking up	Locations	Locations	Sit reports	Grid refs	Locations
	NAIs	Minefield				Javelin	POWs	Sit reports	Marking up	Likely enemy positions	BMDs	Comms
	Grid refs	Call signs						Kills	FOO	Dobby	Sit reports	Contacts
	Crossing site	Assets							Grid refs	Contacts	Scabbers	
	Bridges	Time							POWs	Kills		
	Phase lines	Battle							River	Crossing		
									Intent	Remaining Platoon		
									Platoons	Recce vehicles		

Note: Shaded items denote information elements that were transacted across phase lines battle execution process.

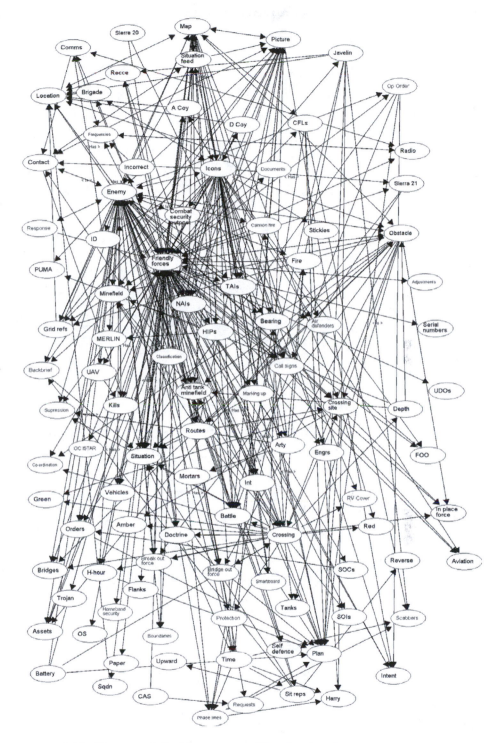

Figure 8.19 Harry and Scabbers propositional network

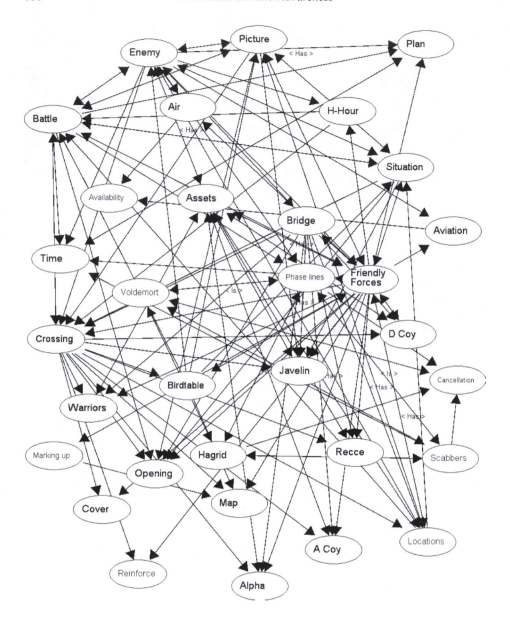

Figure 8.20   Voldemort propositional network

The shaded items within Table 8.2 represent those information elements that were
exchanged across the different phases of the battle (i.e. which were key information
elements during each phase line enactment). The key information elements are useful
in this case as they represent the key pieces of information that were critical to the
battle execution process. In this case, it is notable that there were issues surrounding
the accuracy and timeliness of the presentation of a number of the key information

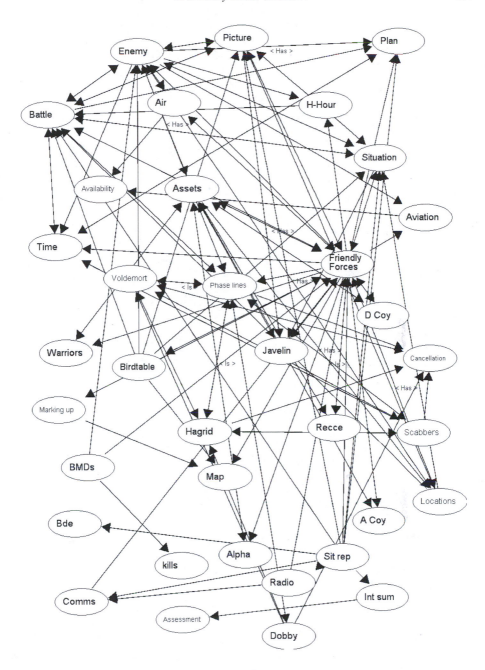

Figure 8.21   Hagrid propositional network

elements. For example, enemy and friendly force location information was typically not presented in a timely fashion and so was often not compatible with the actual state of the world.

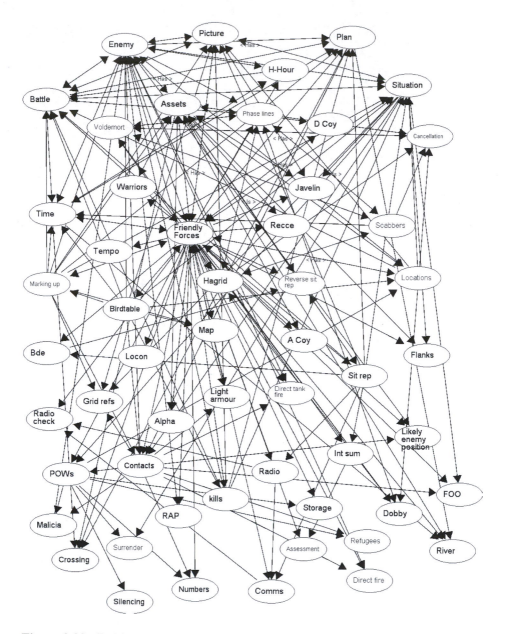

Figure 8.22   Dobby propositional network

Discussion

The purpose of this chapter was to present an analysis of a new digital mission planning system during the operational field trial exercise. The main aims of the study described were to analyse DSA during land warfare mission planning and execution activities and

Figure 8.23 Dumbledore propositional network

to evaluate the new digital mission support system in terms of its impact on DSA during the activities observed and its support for DSA requirements. It was also intended that the analyses would inform the development of guidelines for the design of collaborative systems. The findings are therefore discussed with regard to three key areas: the nature

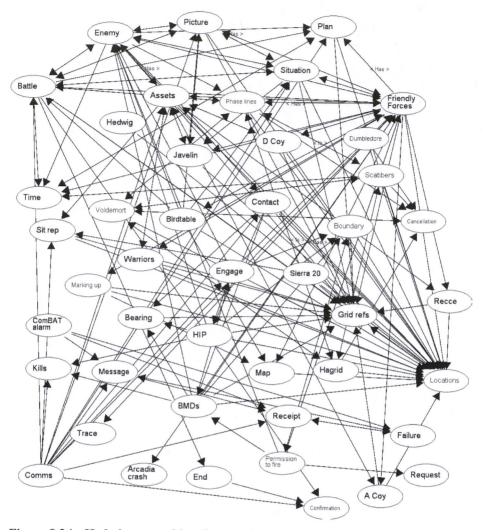

Figure 8.24   Hedwig propositional network

of DSA during land warfare mission planning and execution, the impact of the digital system on DSA and the resultant implications for future collaborative system design.

Distributed Situation Awareness During Mission Planning and Execution

The analysis revealed a number of interesting facets associated with DSA during the mission planning and execution activities analysed. From a theoretical viewpoint, it was notable that the notion of compatible, rather than shared, SA was apparent during the planning and battle execution activities. The propositional networks indicate that the information elements underlying the system's DSA represented compatible, rather than shared or common SA requirements. Each team member was using this information for their own means and the distinct roles and responsibilities extant throughout the planning and battle execution process were such that common or shared SA was neither possible nor would it have been productive. This finding was corroborated by the SA requirements analysis findings, which suggested that each team member had distinct SA requirements. The corollary of this was that, even when different staff were using the same information, they were using it for different purposes and so their SA was different from each other.

The analysis presented demonstrates how, during the planning process, the planning team is divided into distinct cells, each with their own specific role and subsequent goals and tasks to fulfill. For example, during Question 1 (the battlefield area evaluation phase), the engineer cell is primarily concerned with the ground, friendly and enemy forces' use of the ground and the impact that it is likely to have on friendly and enemy operations, whereas the intelligence cell is primarily concerned with the threat posed by the enemy, including their capability, strengths and weaknesses, and enemy doctrine. Thus, even when both team members have access to the same information regarding the battlefield area and the enemy, they use and view the information in a very different manner; it is the relationship between concepts that makes up their distinct SA. Indeed, thinking about SA as the relationship between concepts is the key to the DSA approach; even when team members have access to the same information, the relationships between the information elements is likely to be different based on how they are using the information and what they need it for. In the example cited above, the relationships between the enemy and the battlefield area are viewed very differently by the engineer and the intelligence components; the engineer looks at how the ground may shape enemy operations whereas the intelligence cell looks at the ground and the resultant threat level imposed by the enemy. It is this unique combination of information elements by each team member that makes their SA compatible and not shared. The engineer versus intelligence components views on the enemy and ground during Question 1 are represented in Figure 8.25.

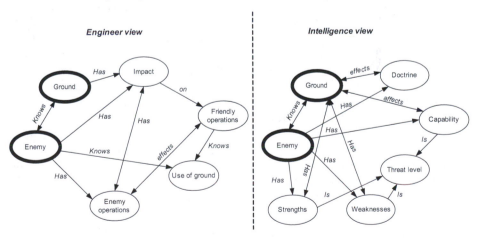

Figure 8.25   Engineer versus intelligence components differing views on enemy and ground information elements

Note: The figure demonstrates how each component is using the information for their own ends and how its subsequent combination with other information makes each view unique.

This clear differentiation between roles and responsibilities brings with it distinctive SA requirements for each cell. It is therefore concluded that the organisation of the teams involved and the presence of specialised roles was such that the majority of the SA elements represented distinct compatible SA elements; each individual and sub-team had their own unique combination of SA requirements depending on the role, goals and tasks that they were required to undertake. This finding is encouraging since it provides further naturalistic evidence to support the notion of compatible SA that is postulated by the DSA theory advocated in this book.

The compatibility of team member SA requirements during the planning and battle execution activities observed has very clear implications for the design of any system intended to support them (and for collaborative systems in which team member roles are distinct). The implication of this is that collaborative system design should be driven with a very clear specification of the compatible SA requirements of the different users; the system should then be tailored to support these unique SA requirements. This permits a system that has the capability to present only the required information to the right users at the right time, a provision that is key for DSA. Rather than simply design a system that presents all the information available (requiring users to locate the information that they require) and contains all the planning tools and functions required by the overall group (requiring users to locate the functions that they require), the tool should instead be tailored specifically to support each role (in terms of information and tools required). User-tailored systems such as this would minimise the overload of agents with unwanted information and tools.

This conclusion is corroborated by other findings presented in the literature. For example, Bolstad et al. (2002) analysed the SA requirements of a US Army Brigade and found explicit differences between the SA requirements of the different officers. In conclusion, they suggested that in military planning systems team members do not need to know everything that the other team members know and that a single display would not meet the needs of all of the brigade officers. Subsequently Bolstad et al. (2002) recommended that in order to provide only the level of detail required for a particular user without presenting unnecessary information, displays should be tailored to each officer's needs whilst also providing information relating to the SA of the other officers in the team. Gorman et al. (2006) also suggested that, due to the specialised roles apparent within typical command and control environments, the design principle of giving every team member displays which present all the information required by the entire team is invalid. Gorman et al. (2006) proposed that it may in fact be prohibitive and counteractive to give everyone mutual access to the same information. Similarly, Kuper and Giuerelli (2007) postulate that in order to enhance command and control team efficiency, tailored work aids should be used to reduce the cognitive load associated with mining through redundant information. They argued that the key to efficient and effective command and control team performance is the design of work aids that support both holistic work practices and unique first person perspectives.

The present analysis indicated that, despite the presence of such explicitly compatible, rather than shared, SA requirements, this has not been taken into account in the design of the mission support system. In its current form, the digital system does not support the compatible SA requirements of its different users. Rather, the system simply provides the same displays, tools, interface and, more importantly, information to every user regardless of roles and goals. It is not customisable nor can it be tailored based on different user requirements. The onus is thus placed on the user to find the information and tools that they need within the system, a process which is time consuming and difficult, particularly for new users.

Digital Systems Impact on Distributed Situation Awareness

The analysis also provided compelling evidence of the impact on DSA that the digital system had during the activities observed; these findings can be discussed with regard to the hypotheses set out at the beginning of this chapter. A judgement of whether or not the digitised mission support system met the requirements stipulated at the beginning of this chapter is presented in Table 8.3.

The first and perhaps most telling finding was that, in undertaking the required activities the teams involved continued to use the traditional paper map processes in order to support and supplement the new digitally supported process. On the majority of occasions, this was because of flaws present within the new digital system that were adversely affecting DSA. It was simply quicker and easier for users to generate and maintain the level of DSA required using the old paper processes. It was concluded that the mission support system did not adequately support the acquisition and maintenance of DSA during the activities observed; rather, a combination of the paper map process and the new digital systems was used.

Secondly, there were many instances in which the SA-related information presented by the digital mission support system was in fact inaccurate and was

Table 8.3   Digitised mission support system analysis findings

Requirement	Finding	Summary
The system should support the range of different user SA requirements	X	The compatible SA requirements of the different users are not supported; the system is the same regardless of who is using it and is not sufficiently customisable
The system should present SA related information in a timely manner	X	SA-related information presented on the LOP was often out of date (sometimes up to 20 minutes) due to data bandwidth issues
The system should present only accurate SA related information	X	SA-related information presented on the LOP was often inaccurate (e.g. contact reports, positional data)
Users should trust the SA related information presented by the system	X	Due to inaccurate and untimely information presentation issues and usability problems, the users reported that often they did not trust information presented on the LOP or in planning products
The system should support the level of DSA required for mission planning and execution activities	X	DSA was acquired and maintained using a combination of the new digital system and the existing paper map process

not compatible with the real state of the world at the time when it was presented. This is represented in the summary propositional network presented in Figure 8.26. Within Figure 8.26, the darker shaded information elements represent those that were presented inaccurately by the system during the operations observed.

This was particularly problematic during battle execution, where the information presented on the LOP was either out of date or spurious. This meant that the Brigade's and BG's understanding of enemy and friendly force locations, movements, number and capabilities was often inaccurate. To overcome this, radio voice communications were used to supplement and/or clarify contact reports and a paper map with sticky icons was used to represent the battle. These mismatches had the impact of reducing the accuracy of DSA and also adding time to the planning and execution; the result of this was a reduction (rather than the projected increase) in operational tempo.

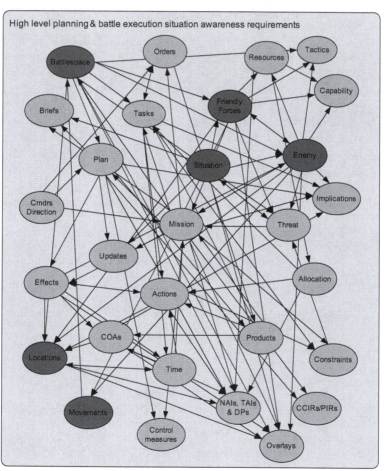

Figure 8.26   Inaccurate information elements

Note: Darkly shaded nodes represent the information that was presented inaccurately by the digital system.

Thirdly, the timeliness of the SA-related information presented by the digital system was also problematic; due to data bandwidth limitations voice transmission was given precedence over global positioning data regarding the locations and movements of entities on the battlefield. Because of this, contact reports and positional information presented on the LOP was often out of date (on occasions being presented up to 20 minutes late). This is represented in Figure 8.27.

The corollary of this was that the system's DSA was at times 'out-of-date' or at least lagging behind the real state of the world. The problem of 'delayed' SA information presented by the system was a bandwidth issue. Specifically, because of the amount of data being transmitted and the limited bandwidth of the system, the voice communications data takes precedence over the OSPR data. This meant that during complex operations the OSPR data is delayed due to high voice communications traffic. Due to the same data transmission problems the digital system was also observed to be slow in updating the enemy positions on the LOP.

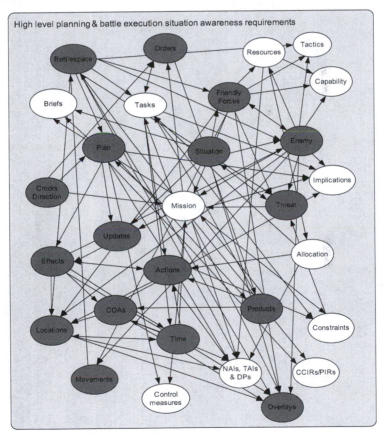

Figure 8.27   Untimely information elements

Note: Darkly shaded nodes represent the information that was not presented in a timely manner by the digital system.

As a consequence of the problems discussed above, a fourth issue identified was the low level of trust that the users placed in the SA-related information presented to them by the digital system. This is represented in Figure 8.28, where the darker shaded information elements represent the information that users of the system did not fully trust during the activities observed (based on discussions with the users and also observation of the activities in question).

During both planning and execution activities (mainly execution) the issue of user mistrust in the SA-related information presented by the system was evident. Mistrust

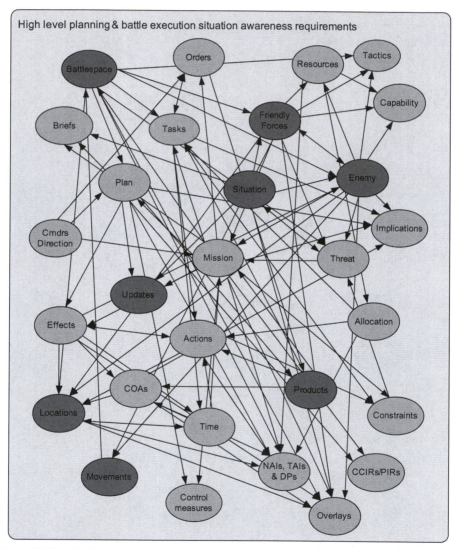

Figure 8.28   Lack of trust in information elements

Note: Those information elements shaded dark grey represent the information presented by the system that users felt was untrustworthy.

in information can adversely affect SA development and maintenance. According to Endsley and Jones (1997) in the event of uncertain information, individuals either search for more information or act on uncertain information, both of which can be detrimental to SA. It is apparent from the analyses that user trust in the information presented by the digital system was minimal; due to issues such as SA mismatches, spurious data and significant delays in the presentation of positional and contact reports many users often questioned the information presented by the digital system and often took measures to clarify the accuracy of the information (e.g. requests for clarification of location and status reports). This served to add to the planning and execution process and also had an adverse impact on the tempo of operations.

A fifth and final issue related to the granularity of the maps used (within the digital system) and their impact on DSA was identified. One of the key issues related to the development of SA of the ground during the planning process that was observed consistently throughout the exercise was the problems with the granularity of the maps used on the digital system. Users found it extremely difficult to analyse the ground and appreciate what was going on when looking at the maps presented by digital system. This meant that the users could not assess the ground sufficiently and, in addition, the size of the display meant that users could not get an overview of the entire battlefield. The only way in which users could see the entire battlefield area was to zoom out; however, this often led to the users losing context in terms of exactly which area of the battlefield area they were looking at. This problem was consistently reported by users throughout the exercise.

Overall, our analysis suggests that the digital system did not provide adequate support for DSA development and maintenance during the planning and execution activities observed. Rather, it was a combination of the digital system and the traditional paper map process that enabled the system to develop and maintain the level of DSA required for successful completion of planning and execution activities. Further, flaws present in the digital system had the effect of degrading DSA and reducing operational tempo.

Further analysis of the system can be informed by the literature on SA and command and control systems. In conclusion to an analysis of the manoeuvres planning process in land-battle situations, Riley et al. (2006) identified the following key issues for effective planning in command and control operations that require support through new technologies and system designs:

1. rapid development and dissemination of plans;
2. visualisation of plans and tracking deviations to planned activities;
3. contingency planning;
4. distributed collaborative planning; and
5. plan rehearsal.

It is possible, based on this analysis, to make judgements on how well the digital mission planning system satisfied each of the requirements described by Riley et al. (2006), who suggested that planning products must be disseminated to appropriate units in a timely fashion. Unfortunately, our findings indicate that the process of developing and disseminating plans was significantly lengthened due to problems with the new system. These problems included usability problems with the various tools, data transfer problems and problems with the printing of planning products. It was in fact concluded from the overall analysis that the planning process was significantly lengthened due to the new digital system. Riley et al. (2006) also articulated the need for rapid visualisation and comprehension of plans, the requirements for which included tools to present visualisation of unambiguous elements of the battle (e.g. terrain and weather, weapons and sensor capabilities etc.), tools to support global SA and also tools to support the understanding of deviations between what was planned and what is really happening on the battlefield (comparisons of plan components against battle states). The digital system analysed does not currently have a global SA function or a comparison of planned versus battle states function and so team members did not have the capability to make accurate comparisons between current battle state and the corresponding stages of the plan. The capability for contingency planning refers to the ability for planners to incorporate contingency plans at various decision points within the proposed courses of action (Riley et al., 2006) and also for planners to be able to modify plans quickly and easily. According to Riley et al. (2006) such plans should explicitly define the triggers (e.g. enemy locations, level of combat effectiveness etc.) that indicate when the need for a particular contingency arises. Unfortunately, although the system analysed in this case does permit planners to plan 'on-the-fly', so to speak, the problems associated with the usability of the tools and the length of time that it takes to produce and disseminate planning products limits the efficacy of this function. Currently it takes too long to re-plan and disseminate revised planning products. On a positive note, the digital system did support distributed collaborative planning, since it provides a secure voice network, information sharing facilities and a digital messaging facility. The final key issue cited by Riley et al. (2006) was the need for virtual or simulated rehearsal functions which could enable courses of action to be compared, evaluated and refined accordingly. Currently the digital system does not provide this functionality and during the activities observed simulated rehearsal was achieved via wargaming using paper maps and stickies.

Implications for System Design

The findings derived from this analysis have clear implications for the design of software tools to support collaborative activities. In particular, the following guidance emerges from this analysis:

- *Clear definition and specification of SA (or information) requirements.* The findings suggest that the collaborative system design process should begin with a clear definition and specification of the SA requirements of the different users of the system in question. This should include a description of the process involved, the different roles and tasks involved in the process and a description of who needs to know what and when in the process they need to know it. Clearly, the designers of the system in this case did not fully appreciate the distinct roles and SA requirements of the different end users. Although this principle sounds somewhat obvious, unfortunately it is not always adhered to.

Matthews et al. (2004) point out that knowing what the SA requirements are for a given domain provides engineers and technology developers with a basis to develop optimal system designs to maximise human performance rather than overloading workers and degrading their performance. Matthews et al. (2004) suggest that, 'it is important, therefore, to know the SA requirements for various jobs to design systems that optimally present information, to evaluate the impact of new technology, and to develop effective training procedures to prepare workers to interact with advanced information systems' (p. 160). Matthews et al. (2004) also suggest that 'systematically identifying what it is the worker needs to know to accomplish key goals is a fundamental step in designing technological systems that optimise work performance' (p. 161) and that SA requirements analyses findings can be used to develop appropriate measures of SA for assessing the final system in terms of its support for SA requirements.

- *Design system to support compatible SA requirements.* The findings suggest collaborative systems should be designed to cater for the compatible SA requirements of its end users. Within collaborative systems, users more often than not have distinct SA requirements and so the system should be designed so that users are not presented with information, tools and functionality that they do not explicitly require. The system should therefore be designed to support the roles, goals and SA requirements of each of the different users involved in the process. This might involve the provision of different displays, tools and functions for the different roles and tasks involved. This removes the problem of high workload and getting bogged down in too much data and also reduces the requirement to send large products and data sets to every agent working within the system. In the same way that everyday PCs can be adapted by users so that the user interface and its functionality suit their own needs, it may also be more appropriate to allow the system and interface to be customisable based on the user's role (e.g. G2) or on the job that the user is working on at a particular time (e.g. synch matrix) which will remove the vast number of redundant components of the system that get in the way when the user is doing his or her specific job. Gorman et al. (2006) advocate adaptive and timely information sharing, which they stress does not mean that everybody has access to the same information at the same time, but rather implies communicating the appropriate information (and, importantly, no more than this) to the right person at the right time. In this case, the analysis indicated that the distinct SA requirements of the different end users were not supported in any way; rather the system remained the same in terms of information presentation, interfaces, tools available and functionality regardless of who was using it. The principle of providing system elements only with the information that they require becomes even more critical with the advent of NEC systems, where the great increases in information communicated around the system mean there is considerable potential for informational or data overload.
- *Use multiple interlinked systems for multiple roles and goals.* When a team is divided into distinct roles and team members have very different goals and informational requirements it may be pertinent to offer separate (but linked)

systems. In the same way that Microsoft Office provides separate word processing (e.g. Word), drawing (e.g. Visio) and spreadsheet (e.g. Excel) tools, distributed team working support systems should provide a suite of mission support tools catering for the different users and roles involved; each tool should have the functionality and information required for the role it is designed to support whilst also containing the ability to see global information. As referred to above the system focused on in the current study remained the same regardless of who was using it.

- *Customisable/tailored interfaces.* As articulated previously, the nature of collaborative systems is such that there are specific roles and SA requirements. Subsequently, the information and the tools that one agent needs to use may be very different to those that another agent needs. Collaborative systems should therefore be customisable, allowing users to customise (either by them or intelligently by the system based on usage) the interface so that the information and tools that they specifically require are present. This increases the usability and ease of use of the system and also reduces interaction time (i.e. having to mine through menus to find information and tools required).

- *Consider technological capability and impact on DSA.* Again, perhaps an obvious, but nevertheless critical, recommendation is that system designers need to consider carefully the constraints imposed on them by technological capability and design the system accordingly. DSA in this analysis was adversely affected by both the capability of the displays and mapping used and also by bandwidth limitations. It is therefore recommended that systems be designed within the constraints of the technology available.

- *Ensure the accuracy of information presented.* It goes without saying that the information presented by any command and control system should be highly accurate. System designers need to ensure that the information presented by all aspects of the system is accurate at all times. The present study revealed that the mission support system under analysis did not always present accurate SA-related information such as contact and positional reports and enemy and friendly movements on the battlefield; further, this information was often not presented in a timely manner.

- *Design for tempo.* Complex collaborative systems are typically used to support time-critical activities. It is therefore crucial that such systems are designed to enhance rather than inhibit operational tempo.

- *Provide filtering functions.* When systems have displays containing movement and location information relating to distinct entities (e.g. enemy, friendly, neutral etc.) on a map, it is important that the system allows the users to filter the display so that different classes of information only are displayed.

- *Clear communications links.* Throughout this research the importance of communications links for DSA acquisition and maintenance has consistently been highlighted; additionally a number of other researchers have identified communication links as key to team SA (e.g. Gorman et al., 2006; Stanton et al., 2006; Walker et al., 2006) It is therefore critical that collaborative systems posses the appropriate communications links and that the users working with the system understand which communications channels are and are not open to them and also understand when and to whom what information should be communicated. This follows on from Stanton et al.'s (2006) conclusion that the links between agents in a network are at least as important as the agents themselves in maintaining DSA.

- *Test DSA throughout the design lifecycle*. It is clear that DSA should be considered and tested where possible throughout the design lifecycle. DSA requirements should be used to drive the design of concepts and concepts should be evaluated based on their ability to meet the DSA requirements of the end-users.

In closing it is the authors' opinion that, in isolation, the digital system analysed did not provide adequate support for DSA during planning and execution activities. Rather, it was a combination of the digital system and the analogue paper map process that enabled the system to develop and maintain the level of DSA required for successful completion of the planning and execution activities observed. Although the digital system does appear to have the potential to support DSA during planning and execution activities, at present it falls short of this key requirement in a number of areas and consequently, a combination of both planning systems (digital and analogue) was used throughout the activities observed. The issues limiting the level of DSA afforded by the new digital system included the timeliness and accuracy of the information presented and the presentation of appropriate information to the appropriate users, all of which subsequently affect the trust that the users place in the SA-related information presented by the system. On a positive note, the system does appear to present the information required for DSA and also provides the communication links required for DSA to percolate throughout the system.

The huge potential of digitising warfare systems and processes can only be realised with further investigation and evaluation in order to determine how systems can be better designed in order to enhance DSA and ultimately mission planning and execution activities. Key issues to pursue relating to the concept of SA include what information should be presented, in what manner and to which elements of the warfare system, how information can be presented in a more timely fashion and how the accuracy of information presented by command and control systems can be enhanced and ensured. Ultimately the great potential that digitisation offers for enhancing mission planning and execution activities in the military domain is also accompanied by a very real opportunity to create warfare systems in which activities become more difficult and complex, more prone to error and subsequently less efficient.

Chapter 9
A Model of Distributed Situation Awareness in Complex Collaborative Environments

Introduction

The purpose of this research was to explore and extend the theoretical foundations for a DSA proto-theory laid by Stanton et al. (2006) and to investigate the concept further in terms of its measurement and its implications for collaborative system, training programme and procedure design. In order to extend Stanton et al.'s (2006) theoretical foundations the final phase of the research involved developing a prototype model of DSA in complex collaborative environments based on the findings derived up to this point. In particular, an explanation of how DSA functions in complex collaborative environments, along with the factors affecting it, was required. This chapter presents a model of DSA in complex collaborative environments that was formulated based on the findings of the research up to this point.

Distributed Situation Awareness Model

The review of SA models presented in Chapter 2 suggested that there is currently a lack of a model of SA in collaborative environments that fully describes the processes involved, the content of a system's DSA and also the factors affecting SA. In response to this it was suggested that DSA approaches are more appropriate than existing team SA models; however, it was also noted that comprehensive models of DSA currently do not exist. Stanton et al. (2006) laid the foundations for a model of DSA by proposing a series of tenets of DSA (see Chapter 4), but did not go as far as presenting a complete model of DSA in complex collaborative environments.

The overall aim of this research was to extend Stanton et al.'s (2006) model in order to explain more fully the concept of DSA. To satisfy this requirement a model of DSA was constructed based on the findings derived from the case studies undertaken during this research. In presenting a model of DSA the intention is to, using the findings derived from the research undertaken so far, attempt to describe how DSA functions in collaborative systems and attempt to describe the various factors that are likely to impact DSA acquisition and maintenance. In doing so, the model brings together four strands of thinking: schema theory; the perceptual cycle model of SA and the concepts of compatible and transactive SA. The model of DSA in complex collaborative environments is presented in Figure 9.1.

Using the distributed cognition and cognitive systems engineering perspectives described in Chapter 2 and building on the account of SA presented by Smith and

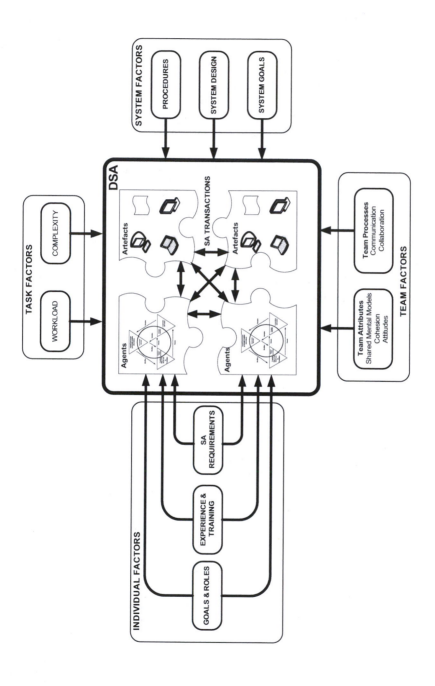

Figure 9.1 Model of distributed situation awareness in complex collaborative systems

Hancock (1995) and the DSA description provided Stanton et al. (2006), the model presented in Figure 9.1 uses schema theory as its basis and treats DSA in collaborative environments as a systemic property that emerges from the interactions (referred to as *SA transactions*) between system elements (both human and non-human). DSA is viewed as the system's collective knowledge regarding a situation that comprises each elements compatible awareness of that situation. SA in collaborative environments is therefore viewed not as a shared awareness of the situation by different team members (e.g. Endsley, 1989, 1995a; Endsley and Robertson, 2000) but rather as the system's collective awareness of the situation comprising each element's compatible portion of SA required for task performance.

According to the model, systemic elements each hold the information that the overall system requires for the development and maintenance of DSA during task performance and this information is passed around the system as and when required via SA 'transactions' that take place between the elements. SA transactions refer to the exchange of SA-related information between system elements and can include communications between elements (explicit and implicit) and interactions with devices (e.g. checking a display). The system's DSA and each individual's SA is dynamically maintained and updated via these transactions. Each systemic element therefore holds a portion of the SA that is critical not only to its own task, goals and roles but also to the entire system's DSA and overall performance. Whilst this awareness might often be built on the same pieces of information, it is not shared SA since each element views it differently based on goals, roles and tasks being undertaken and also experience, training and the resultant schema. Since SA comprises concepts and the relationships between them, each agent's SA is different. Each agent's awareness is therefore compatible with every other's in that it is different to other team member's SA but is collectively required for the system to achieve its desired aim.

DSA Mechanisms: Schema Theory, the Perceptual Cycle, Compatible SA and SA Transactions

The model of DSA presented in Figure 9.1 presents a high-level view of DSA and the factors affecting it in complex collaborative systems. To explain the model further, this section focuses on the mechanisms underlying a system's DSA. There are four key mechanisms underlying the model presented in Figure 9.1, namely *schema theory, the perceptual cycle, compatible SA* and *transactive SA*. These concepts are discussed in more detail below.

Schema Theory and SA

Introduction to Schema Theory

The model of DSA presented uses schema theory as its basis for how the individuals working within a system develop and maintain their SA. Schema theory first emerged in the early 1900s (e.g. Head, 1920; Piaget, 1926) and describes how individuals possess mental templates of past experiences which are mapped with information in the world to produce appropriate behaviour. Bartlett (1932) introduced the concept of 'schema' as active organisations of past reactions and past experiences, which are combined with information in the world in order to produce behaviour. A schema, therefore, is rather like a form of mental template; it is clearly 'more than a "set" because it is more elaborate and less restricted to a particular situation; it is more ideational or implicit than a "strategy" and conceptually richer than a "hypothesis"' (Reber, 1995, p. 689). Bartlett (1932) used the example of cricket to demonstrate how, when making a stroke, a batsman is not producing entirely new behaviour, nor is he merely repeating old behaviour. Rather, Bartlett (1932) suggests that the stroke is 'literally manufactured out of the living visual and postural "schemata" of the moment and their interrelations' (Bartlett, 1932, p. 201). Bartlett's example demonstrates how schemata in the mind of the individual combine with their goals, the tools that they are using and the actual situation in which they are placed in order to generate behaviour. Bartlett (1932) further investigated the concept and the role of schema in an individual's recall of events by undertaking a series of studies on the processes of remembering and forgetting. In conclusion, Bartlett (1932) argued that literal recall was very rare and, rather, that recall was a process of reconstruction and that memories showed evidence of consolidation, elaboration and invention, using material from other schemata.

Bartlett subsequently argued that schema allow individuals to orientate themselves towards incoming stimuli and adapt their responses to it. This frame of reference can work to the advantage or disadvantage of the individual. If the schema is appropriate to the situation, then an appropriate response may be produced. Norman (1981), however, has suggested that the 'triggers' of the situation may be wrongly interpreted, leading to a maladaptive response. Schemata are not necessarily open to conscious examination, so the question of identification and adaptation of more appropriate schema is a moot point. The schema themselves are unlikely to exist as separate sets of templates, but rather as an interconnecting set of structures, aspects of which are triggered in response to a particular set of circumstances or experiences. Thus we could view the activated aspects of schemata as structures that move in and out of pre-conscious (and possibly conscious) attention like the brightening and dimming of variable lighting. Neisser (1976) suggests a hierarchical arrangement of embedded schemata and their associated actions. As proposed by schema theorists (Bartlett, Piaget, Neisser and Norman), the schemata are continually modified through interaction with the world in which behaviour is created.

Anderson (1977) suggested that there are five main defining features of schemata. These include that the schemata: 1) are organised meaningfully in some way; 2) are embedded within other schemata and contain sub-schema themselves; 3) change from moment to moment as information is received 4) are re-organised when incoming data reveals a need to restructure; and 5) are gestalt mental representations. These features allude to the dynamic, non-linear and personal nature of schema, which is why Bartlett noted that memories of events (even learning of stories) take on such an individual nature when recalled. They also account for the performance differences between novices and experts, as experts might not only be attending to different stimuli (as directed by their schemata), but also deriving different types of understanding through their interaction. Further, the gestalt nature of the schema could mean that experts are able to infer more than simply the bare facts might suggest, implying that a higher level of understanding can be derived though richer schema and interactions.

Norman and Shallice (1986) used the ideas behind schema theory to develop a cognitive model of attention and control that could be used to explain everyday behaviour. They distinguished between automatic and willed control and argued that schemata are templates for behaviour that are triggered by cues in the environment. Although several schemata might be activated at any moment in time (offering a range and variety of possible behaviours), the selected schema will be automatically allocated on the basis of the strength of activation and motivations of the individual. Controlled processes are only activated when the task becomes too difficult, such as novel situations or when errors are made.

Genotype and Phenotype Schema

Baber and Stanton (2002) describe the concepts of global prototypical routines (GPRs) and local state specific routines (LSSRs) in order to explain how individuals interact with products and devices. They suggest that individuals use GPRs and LSSRs to direct their interactions with products and devices and that GPRs represent the schemata in the mind of the person whereas LSSRs represent the activated schema brought to bear on a specific problem by a user. Similarly to Bartlett (1932), they suggest that the schema is reconstructed with the current stimuli and that the ensuing interaction leads to the modification of the schemata toward the goals (although even the goals are subject to change in light of the interaction). GPRs represent stereotypical responses to system images that a person has learned, acquired or otherwise developed (Baber and Stanton, 2002). Examples of GPRs include a strong stereotyped response to turn a tap (faucet) anti-clockwise to turn it on or to increase water flow (Sanders and McCormick 1992; cited in Baber and Stanton, 2002). Regardless of whether these responses are correct it is important to note that individuals typically attempt them before any other actions (Baber and Stanton, 2002). Baber and Stanton (2002) also propose that individuals possess LSSRs which involve the generation of appropriate actions through the individual's interpretation of a device's 'system image' in relation to the current goal state (Baber and Stanton, 2002). LSSRs are therefore dependent on the information that is available through the system image. Baber and Stanton (2002) suggested that designers of those public technologies that expect people to use them accurately the first time, without any instruction, such as food vending machines, ticket machines and automated teller machines, need to capitalise on triggering appropriate GPRs and supporting the user in adapting LSSRs.

GPRs are rather like the genotype schemata and LSSRs are rather like the phenotype schemata proposed by Neisser (1976). Genotype, in this context at least, refers to the wider systemic factors that influence the development of individual cognitive phenomena and behaviour. The local, individual-specific manifestation of cognition and behaviour represents the phenotype. As part of the theory underlying his Cognitive Reliability Error Analysis Method (Hollnagel, 1998), Hollnagel (1998) uses the genotype and phenotype distinction to illustrate how generic error modes (the genotype) and may be related to observed errors (the phenotype) in the world. Hollnagel's model suggests that the combination of genotypes (man, technological and organisational), the environment and random variation produces the phenotype, which is the observable manifestation of the error.

It is apparent that, more often than not, devices fail to trigger appropriate schema in their users. Norman (1981) used schema theory to explain erroneous actions such as slips of action or lapses in attention. His analysis suggested that three basic genotype schema-related errors could account for the majority of errors. These were activation of wrong schemata (due to similar trigger conditions), failure to activate appropriate schemata (due to a failure to pick up on the trigger conditions indicating a change in the situation) and a faulty triggering of active schemata (triggering the schemata either too early or too late to be useful).

Neisser's Perceptual Cycle

Neisser's (1976) seminal work *Cognition and Reality* is perhaps the most commonly used and cited text on schema. Here Neisser described the concept of the perception-action cycle, which included the notion that anticipatory schema held by individuals served to anticipate perception and direct action. Neisser proposed the ecological view in juxtaposition to the information processing view. The ecological approach suggested that perception was an active, rather than a passive, process and that perception could be viewed as guided exploration in the sense that the active schemata direct where we look/listen/touch and what we expect to see/hear/feel. This exploration leads to adaptation to the environment by the perceiver, which guides future exploration. Neisser adopted the view that interaction with the world is cyclical in nature rather than linear, as implied by an information processing chain. The schemata are the *active knowledge structures* that guide the exploration and interpretation of the information, which in turn changes those structures, further guiding exploration, and so on. The form and nature of the schema will determine what we are able to perceive through this interaction, i.e., how it fits into our own personal schemata. Neisser argued that schema interact with the temporal nature of events by linking the past to the future in two main ways. First, the anticipation of what will happen next determines what we do – what information we look for and attend to. Second, we understand the stream of activity though the anticipation (and continuous modification of that anticipation) of being able to make sense of the events as they unravel through the interaction. We see/hear/feel/smell/taste the whole experience in terms of its meaning to us as individuals.

Neisser considered the multiplicity of information and integration of the modalities essential to the interpretation of the experience. Thus, schema-based theories tacitly assume that cognition is not only cyclical (rather than linear) but also parallel (rather than single channel). The schema are modified by the experience, but themselves are also modifying the experience creating, if you will, a better situation for the individual to be aware of. In this way, Neisser links cognitive activity to physical behaviour to exploration and interaction in the world. To a psychologist, the perception-action cycle together with schema theory offers a theory of everything. It explains the way in which the world constrains behaviour as well as how cognition constrains our perception of the world. It explains both top-down and bottom-up processing of information, but also shows that everyday behaviour is formed through a mixture of both approaches. Whether we process features or meaning is extracted from features depends upon which part of the perceptual cycle we are in, which in turn directs the 'information pick-up' next time around. Hollnagel (1993) proposed the perceptual cycle as a fundamental unit of analysis in the assessment of 'joint cognitive systems', such as can be found in human-computer interaction.

Smith and Hancock (1995) used Neisser's perceptual cycle as inspiration to define situation awareness as '*adaptable, externally-directed, consciousness*' (Smith and Hancock, 1995, p. 135). It is this view that forms the basis for the model of DSA presented in Figure 9.1. The approach fits with a wider, increased level of emphasis placed on the collective behaviour of systems as a whole, as opposed to the behaviour of the individuals working within the system (e.g. Artman and Garbis, 1998; Hutchins, 1995; Hollnagel, 2001; Hollnagel and Woods 1999; Ottino, 2003). The work of Hollnagel (1993) reflects this trend. For example, he notes that the 'unit of analysis' of teamwork has to be higher than the level of the individual. Indeed, Hollnagel's well known 'contextual control model' was used to describe the mode of activity 'the team was in', rather than describing the activities of any of its members. Artman and Garbis (1998) also argue that when considering team performance in complex systems it is necessary to focus on the joint cognitive system as a whole and that, in domains such as the military, teamwork is essential for success. The corollary of this, as Ottino (2003) states, is that complex systems cannot be understood by studying their parts in isolation, rather that the real meaning of the system lies instead in the interaction between its parts and the resultant behaviour that emerges from these interactions. Thus, a non-individual approach to the assessment of SA fits well with wider movements in the literature.

Smith and Hancock (1995) identify SA as a subset of the content of working memory in the mind of the individual (in one sense it is a product). However, they emphasise that attention is externally directed rather than introspective (and thus is contextually linked and dynamic). Relating Smith and Hancock's model to genotype and phenotype schema suggests that individuals possess genotype schema, which are triggered by the task relevant nature of task performance. During task performance, the phenotype schema comes to the fore. Although these genotype and phenotype schema may not be open to analysis, we would argue that it is likely that the phenotype schema may be inferred though a variety of data collection methods. Smith and Hancock argue that the 'unit of analysis' should be at the level of the interaction between agents and artefacts, rather than individual consideration of each separate component. The perceptual cycle offers insight into this interaction as well as defining how agents maintain an awareness of changing situations, on a moment-by-moment basis. Adams et al. (1995) argue that the perceptual-action cycle illustrates how it is possible for people to maintain SA 'provided that the flow of data is manageably paced and reasonably compatible with the knowledge and experience constituting the perceiver's active schema' (Adams et al., 1995, p. 90). It is probable that when the workload is too high to maintain awareness, people revert to genotype schemata, as it may not be possible to maintain the phenotype.

Hole (2007), for example, notes that cognitive theorists (e.g., Norman and Shallice, 1986) propose separate supervisory and scheduling sub-systems that attempt to resolve conflicts in attentional demands.

It is apparent that there is significant incongruence between the ideas of schema-driven SA in collaborative environments and the shared SA view (e.g. Endsley and Robertson, 2000). According to the perceptual-action cycle view, each team member constructs their own personal mental theory of the situation, perception becomes reality and the situation, whatever it may be, is modelled differently by each team member. The role of personalised genotype and phenotype schema further discounts the shared SA view. It is therefore apparent that it is not possible for individual team members to share SA with one another. Individual team members may be using the same information as one another and may even have the same SA requirements as one another; however, variability present in their goals, roles, experiences, training, knowledge, skills and attitudes makes the presence of shared SA between them questionable. The presence of different goals, roles, experiences, training, knowledge, skills and attitudes across team members suggests that each team member's genotype schema will be unique, regardless of whether the information that they are exposed to is identical. Indeed, the findings that emerged throughout this research confirm that SA in collaborative environments is not shared between team members; rather it is compatible. Whilst the information used to construct SA may be shared (as in used by different team members) between team members, the resultant SA is not shared since it is different based on goals, roles, tasks and experience.

Compatible Situation Awareness

The key difference then between existing team SA models (e.g. Endsley and Robertson, 2000; Salas et al., 1995) and the DSA model presented therefore relates to the issue of shared versus compatible SA. The concept of compatible SA takes a different approach to the shared SA view and is based on the notion that no two individuals working within a collaborative system will hold exactly the same perspective on a situation. Compatible SA therefore suggests that, due to factors such as individual roles, goals, tasks, experience, training and schema, each member of a collaborative system has a unique level of SA that is required to satisfy their particular goals. Each team member does not need to know everything, rather they possess the SA that they need for their specific task but are also cognisant of what other team members need to and do know. Although different team members may have access to the same information, their resultant awareness of it is not shared, since the team members often have different goals, roles, experience and tasks (and thus different schema) and so view the situation differently based on these factors. As Salas et al. (1995) point out, an individual's pre-existing knowledge and cognitive processing skills influence their SA. Each team member's SA is, however, compatible since it is different in content but is compatible in that it is all collectively required for the system to perform collaborative tasks successfully.

Sonnenwald et al. (2004) suggest that in most team situations not all team members can, or should, have the same shared understanding of the situation. Therefore, it is argued that to suggest that all team members have their own SA and also shared SA with other team members and the overall team is an oversimplification. Any sharing of goals, intent and understanding arises out of the need of the individual team members to perform their tasks and not for its own sake. The ideas of 'sharing' have mutated into a vague belief that sharing ensures a cohesive team, whereas it seems more appropriate that 'compatibility' will in fact lead to cohesiveness. DSA requirements are thus taken to be different from shared SA requirements (Stanton et al., 2006). According to Stanton et al. (2006) shared SA implies shared requirements and purposes whereas DSA implies different, but potentially compatible, requirements and purposes.

It is worth pointing out that the compatible SA concept does not in any way suggest that there are no longer shared SA requirements (i.e. common SA requirements) across team members. Shared SA requirements in this sense means that different team members may need to 'know about' the same information in order to achieve their goals during task performance. However, this does not mean that when they are using the same information they are sharing awareness as they still have different SA to one another. This notion of compatibility between team members' SA as opposed to shared SA between team members is represented in Figure 9.2.

The concept of compatible SA underlying a system's DSA has been consistently demonstrated through case studies described in this book. For example, the findings derived from the energy distribution case studies (see Chapter 5) highlighted how, in each of the scenarios analysed, the different elements of the system all had different but compatible SA. For example, in the scenarios analysed the COCR operator's SA consisted of a very high-level awareness of the activities required and the activities

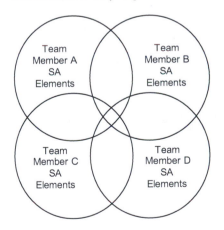

Shared SA (e.g. Endsley & Jones, 2001)

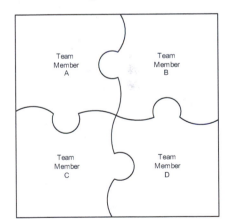

Compatible SA (e.g. Stanton et al., 2006)

Figure 9.2   Shared SA versus compatible SA

being undertaken in the field and also the current status of the ongoing work. This was very different to the agents working in the field (e.g. SAPs and APs), whose SA comprised very low-level specific details relating to the work that they were undertaking at the time. Each knew in generic terms what the other should know but had no specific SA of what they actually did know. The COCR knew what activities the SAP and AP in the field were undertaking and thus what they were aware of, but he did not have a detailed and dynamic SA of their activities. This is not representative of shared SA; rather, although each agent held a different view of the situation, it was compatible with other agents' SA in that each agent's SA formed a composite part of the DSA of the entire network and was required collectively for the entire system to work. Without the COCR operator's SA of the work required, the SAP and AP in the field would not know what work they were required to undertake and similarly, without the SAP's and AP's awareness of their work status and their subsequent transaction of this to the COCR operator the system would not know that its aims had been achieved.

The findings derived from the MNE4 case study also demonstrated the compatible SA concept. These showed how the different sub-groups (e.g. EBP, EBE, EBA, MNIG etc.) had very different SA requirements throughout the activities observed. Further, when SA requirements were common (i.e. the groups were using the same information) the different roles of each group meant that their usage and subsequent awareness of the information was often different.

The studies focusing on land warfare planning and execution activities also highlighted how the different individuals and teams working within the Brigade and BG systems each had unique, very different SA requirements during the planning and execution processes. Each cell within the Brigade and BG system contributed a critical portion of SA that made the entire system work. In addition, a number of common SA requirements were also identified, such as the mission, the commander's intent and the commander's effects. Despite the presence of these common SA requirements, however, the analysis indicated that when presented with the same information, the different elements of the system had very different perspectives on the information presented based on their goals, roles and the tasks that they were required to undertake. Thus, both compatible and common SA requirements were demonstrated. Further, the importance of considering compatible, rather than shared SA requirements when designing collaborative systems was highlighted and one of the key flaws with the digital system was found to be its ignorance of the different SA requirements of its users.

Situation Awareness Transactions

The question remains as to how DSA is built between team members. How do team members 'share' SA if they do not have shared SA requirements? If team members have different SA requirements, then how does the communication of information satisfy each team member's SA requirements? Of course, the compatible SA view does not discount the sharing of information, nor does it discount the notion that different team members have access to the same information; this is where the concept of SA 'transactions' applies. Whilst the concept of compatible SA describes the content of DSA, the concept of transactive SA goes some way to explain how DSA is maintained across the joint cognitive system. The idea of transactive awareness describes the notion that agents within collaborative systems can enhance the awareness of each other' through SA 'transactions'. A transaction in this case represents an exchange of SA information from one agent to another (where agent refers to humans and artefacts). Team members may exchange information with one another (though requests, orders and situation reports); the exchange of information between team members leads to transactions in the SA being passed around; for example, the request for information gives clues to what the other agent is working on. The act of reporting on the status of various elements tells the recipient what the sender is aware of. Both parties are using the information for their own ends, integrated into their own schemata, and reaching an individual interpretation. Thus the transaction is an exchange rather than a sharing of awareness. Each agent's SA (and so the overall DSA) is therefore updated via so-called SA transactions.

Compatible Situation Awareness and Situation Awareness Transactions Example

It is possible to revisit the data in order to demonstrate the concepts described above. The following example is taken from the energy distribution case study described in Chapter 5. Specifically, the example is taken from the return to service scenario, which involved system maintenance and the installation of new equipment.

A propositional network for the scenario is presented in Figure 9.3. Those elements belonging to each of the three sub-teams (i.e. the COCR operator, the SAP and AP working at the substation and the overhead line party working on the overhead lines) and a fourth type where the same element is used by more than one sub-team. The three different codes indicate the compatible elements in DSA, i.e. those elements that are required by each sub-team, that are different to the other sub-teams, but necessary for the system to work. The compatibility of the elements indicates that these elements are not in conflict, rather they indicate the different purposes (and therefore different schemata that will be brought to bear). The fourth category of information element is those transactional elements that pass between sub-teams. As with general systems theoretic principles, the transaction between systems elements implies some sort of conversion of the information received, meaning that information elements will undergo change when they are used by a new part of the system. This change will include the way it is combined with other information elements and the meaning that is applied to it in the context of the goals of the sub-team.

The ownership of the information elements is further explored in Table 9.1; this shows the sub-teams, their tasks and the information elements that they use in pursuit of their goals. The information elements active for the different team roles (shown in the vertical columns) represent the genotypic state of SA at the level of the individual. Where situational elements are matched in the horizontal plane across all team roles then these elements can be regarded as invariants and can be viewed as the genotypic state of 'systemic' SA.

As Table 9.1 shows, there are 60 information tokens in total, 19 of which are transactive (that is, elements that are common to two or more team roles). When people talk of 'shared awareness', they are probably referring to the use of information which they consider to

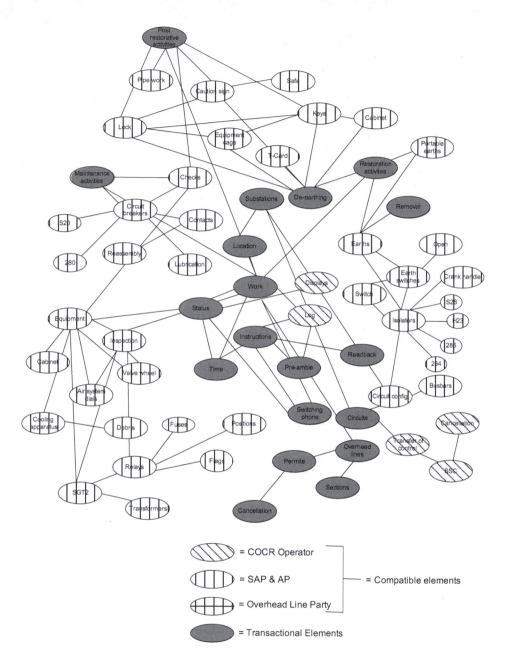

Figure 9.3 **Compatible and transactive elements during return to service scenario**

Table 9.1   **Active situation elements for each team role (the team genotype), the invariants across all team roles (the systemic genotype) and the various 'transactions' between team roles**

	SAP/AP	COCR	Overhead Line Party	
Substations	Transaction	Transaction		Substations
Work	Transaction	Transaction		Work
Location	Transaction	Transaction		Location
Instructions	Transaction	Transaction		Instructions
Pre-amble	Transaction	Transaction		Pre-amble
Log				
Status	Transaction	Transaction		Status
Time	Transaction	Transaction		Time
Displays				
Readback	Transaction	Transaction		Readback
Maintenance activities	Transaction	Transaction		Maintenance activities
Circuits	Transaction	Transaction		Circuits
Circuit breakers				
Checks				
S20				
280				
Reassembly				
Lubrication				
Contacts				
Equipment				
Cabinet				
Inspection				
Valve wheel				
Cooling apparatus				
Air system dials				
Valve wheel				
Debris				
SGT2				SYSTEM GENOTYPE
Transformers				
Relays				
Fuses				
Positions				
Flags				
Switching phone				Switching phone
Permits		Transaction		Permits
Cancellation		Transaction		Cancellation
Overhead lines		Transaction		Overhead lines
Sections		Transaction		Sections
Transfer of control				
System state cert				
Circuit configuration				
Busbars				
Isolators (S28, H23, 286, 284)				
Earths				
Earth switches				
Switch				
Open				
Crank handle				
Removal	Transaction	Transaction		Removal
De-earthing	Transaction	Transaction		De-earthing
Restoration	Transaction	Transaction		Restoration
Portable earths				
Post-restorative activities	Transaction	Transaction		Post-restorative activities
Lock				
Keys				
T-card				
Equipment cage				
Safe				
Caution sign				
Pipe work				

SAP/AP Genotype   COCR Genotype   Overhead Line Party Genotype be identical, such as the information relating to the work status or work instructions as identified in this example. However, whilst it is not argued that the information in this case is related to the same work being undertaken, it is apparent that each component of the system places a rather different meaning and understanding on the work status and work instructions. They are using different genotype schemata to interpret the information and producing different phenotype schemata to pick up information and perform activities related to their tasks and goals. Thus we should be talking in terms of compatible and transactive elements within a general framework of DSA.

We can explore the concepts of transactive and compatible SA elements further by looking at the propositional networks and information elements in more detail. For example, Figure 9.4 presents a snapshot of the SA and activities at different points in time in the scenario. On the left hand side of Figure 9.4 the task in question is described. The information networks presented on the right hand side of the figure depict the information elements comprising SA. Within the information network those information elements that represent transactional and compatible SA are identified.

Distribution of Work Instructions

During the distribution of work instructions, the SAP at substation A is given the earth removal instructions by the COCR operator. Initially the SAP and COCR operator take part in a preamble and once the instructions have been issued, the SAP has to read back the instructions to the COCR in order to confirm successful receipt of them. The information elements *preamble* and *readback* are therefore representative of transactive SA elements. All the other elements, including *time, location, instructions* and the circuits involved, are representative of both transactive and compatible SA, since they are discussed in the context of the work instructions (transactions) but are viewed differently by the SAP and COCR due to their different goals.

Removal of Earth

During the performance of the earth removal, the SAP at substation A undertakes the required activities whilst the COCR operator is engaged in other command activities in the control room. All the information elements are therefore compatible, meaning that the SAP and COCR operator had different, but requisite SA during the activities. For example, the COCR operator's SA consisted of a high-level picture of the various activities being undertaken (e.g. who was doing what, what they were doing and why and what they would be doing next), whilst the SAP's SA was related specifically to his activities at the substation. Thus, although each agent held a different view of the situation, it was compatible with the SAP's SA in that each agent's SA formed a composite part of the DSA of the entire network and was required collectively for the entire system to work.

Work Status Reporting

In the final propositional network, the SAP contacts the COCR operator to confirm that he has completed the removal of earth task. The COCR operator thus receives a 'transaction'

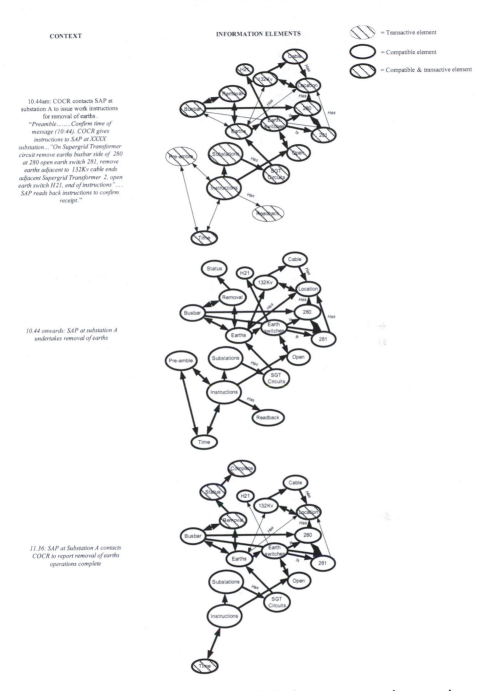

CONTEXT

INFORMATION ELEMENTS

= Transactive element

= Compatible element

= Compatible & transactive element

10.44am: COCR contacts SAP at substation A to issue work instructions for removal of earths.
"Preamble........Confirm time of message (10:44). COCR gives instructions to SAP at XXXX substation..."On Supergrid Transformer circuit remove earths busbar side of 280 at 280 open earth switch 281, remove earths adjacent to 132Kv cable ends adjacent Supergrid Transformer 2, open earth switch H21, end of instructions"....SAP reads back instructions to confirm receipt."

10.44 onwards: SAP at substation A undertakes removal of earths

11.36: SAP at Substation A contacts COCR to report removal of earths operations complete

Figure 9.4   Transactive and compatible SA during return to service scenario

of the SAPs SA via the work status report. In this case then, the elements related to the removal activities (e.g. paperwork, circuit breakers, inspection, lock, earth switches etc.) are representative of compatible SA elements since each SAP has a local and different SA of them at their specific substation. However, the collective awareness of the three SAPs is required for the overall activity to be undertaken successfully. The transactive SA elements during this portion of the task are work progress, time and location, since they are communicated from the SAPs to the COCR operator during work progress updates. This example therefore demonstrates how each agent's SA is different but compatible and also how transactions update the system's SA and serve to prompt further actions.

The COCR operator has SA of the overall ongoing work activities whereas each SAP in the field has SA related to the work that they are undertaking. Each portion of SA is therefore different but compatible and is required collectively for the system to work. These concepts can be demonstrated further by overlaying the energy distribution systems activities onto Smith and Hancock's (1995) perceptual cycle model of SA. This is presented in Figure 9.5.

Figure 9.5 demonstrates how the activities and SA transactions occurring within the energy distribution system can be mapped onto the perceptual cycle model. The first transaction to take place is the issue of instructions by the COCR operator. This serves to update each SAP's schema of the system and of the work required, which in turn drives the activities that the system then undertakes. The outcome of these activities is then checked by the SAPs in the field and the COCR at the control centre (via circuit displays) which in turn modifies both the systems and the SAP's and COCR's schema of the current status

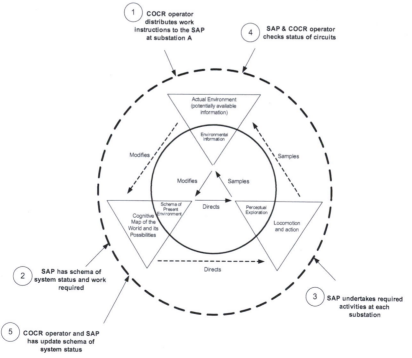

Figure 9.5   System perceptual cycle DSA example

of the system. The examples presented demonstrate how the cyclical perception-action notion can be applied to the entire system as well as the teams and individuals working within it. The COCR and the SAPs involved each initiate SA transactions regarding the state of the environment that serve to initiate action of some sort, which in turn modifies other agents' and the system's schema, which in turn initiates further action and also further transactions regarding the state of the system's schema.

It is also instructive to consider how shared SA approaches would view the same scenario. Endsley's shared SA model (chosen because of its popularity), would typically use goal directed task analysis to identify SA requirements (in the form of SA elements) prior to task performance followed by SAGAT approach to assess team member's perception, comprehension and projection of these SA elements. Subsequent comparisons would then be made on the extent to which team member SA was the same on those SA requirements that were shared. As a starting point in this comparison, it is notable that a SAGAT-based assessment of SA during the scenarios in question was not possible. The scenarios were real world scenarios and so could not be frozen in order to administer queries; nor could queries have been easily administered on-line during task performance. In suggesting how shared SA models would represent this example, one can only assume that a judgement would have to be made on what SA elements were shared and what SA elements were distinct. A representation of how shared SA models may view this example is presented in Figure 9.6.

This example has illustrated some of the basic concepts in the DSA approach. It demonstrates the phenotypical nature of knowledge activation for individual team roles,

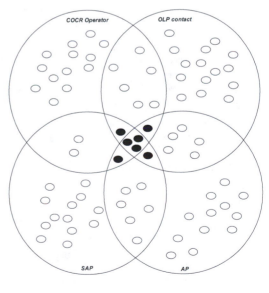

Figure 9.6   Shared SA perspective

that is, their present situational reality and the extent to which that situation differs from other team members'. The example also shows where these phenotypical states have interfaced with other phenotypical states in the form of transactions. Again the point is made that what one element means within one team member's situational model is likely to be quite different from another's. This fact does not, however, diminish its 'compatibility'. Having identified the heterogeneous nature of transactive SA, at the level of the individual, the case study has also highlighted the homogenous nature of SA at the level of the system. Some situational elements are invariants across all actors. Whilst each actor will place their own meaning on these elements and use them for different means, their invariant nature permits diagnosis of the overall genotypic state of a system's SA, as an emergent property of its component/individual states. In principle, this paves the way for diagnosis of how that state is achieved and maintained.

Factors Affecting DSA

The model also encompasses various factors that affect the quality of a system's DSA. These factors can be grouped under the headings of individual, team, task and system factors. Although further exploration is required, each of the factors affecting DSA is discussed briefly in the following section.

Individual Factors

In terms of individual SA this model follows Smith and Hancock's (1995) perceptual cycle model of SA which asserts that SA is a 'generative process of knowledge creation and informed action taking' (Smith and Hancock, 1995, p. 142). According to the model, individual agent SA is developed and maintained through a schema-driven perception-action cyclical process. This is influenced by the agent's goals and roles within the system, experience and training and SA requirements. It follows then that individual factors such as goals and roles, training, experience, schema and SA requirements all have an impact on individual agent SA and subsequently the DSA of the entire system.

The goals of the agents involved are particularly critical, since they are the foundations for each agent's SA as they invoke the relevant schemata and also ultimately impact the way in which they view the situation. Smith and Hancock (1995), for example, suggest that SA is referenced to those goals and boundaries imposed on performance. Schema directs the individual's exploration of the world. An individual's goals therefore plays a part in their interaction with the world and individuals with differing goals may view and use the same information differently in light of their goals. Similarly, an individual's role within a team also has an impact on their SA and subsequently the system's SA. For example, a military commander has very different goals to an infantry soldier and their SA differs accordingly.

An individual's experience and training also affect SA acquisition and maintenance since they both serve to build up and develop schema. Experience of tasks and situations in the form of internally held schema are particularly relevant to an operator's SA.

According to Neisser's (1976) perceptual cycle model (on which Smith and Hancock's model of SA is based) a person's interaction with the world (termed *explorations*) is directed by internally held schemata which are based on previous experiences of the world. The outcome of interactions modifies the original schemata, which in turn directs further exploration. This process of directed interaction and modification continues in an infinite cyclical nature. It is therefore clear that an individual's experience moulds their schema, which ultimately determines the way in which they view the world.

Team Factors

Team attributes and processes are obviously a critical factor in the development and maintenance of a system's DSA. Teamwork is formally defined by Wilson et al. (2007) as "a multidimensional, dynamic construct that refers to a set of interrelated cognitions, behaviours and attitudes that occur as team members perform a task that results in a coordinated and synchronised collective action" (p. 5). A number of researchers have attempted to describe the various processes underlying teamwork. At a simplistic level, team activity can be divided into two forms of behaviour: teamwork and taskwork. Teamwork refers to those instances where individuals interact or coordinate behaviour in order to achieve tasks that are important to the team's goals (i.e. behavioural, attitudinal and cognitive responses coordinated with fellow team members), whilst taskwork (i.e. task-oriented skills) describes those instances where team members are performing individual tasks (in light of their individual roles within the team) separate from their team counterparts. According to Glickman et al. (1987; cited in Burke, 2004) team tasks require a combination of taskwork and teamwork skills in order to be completed effectively.

Research into teams has led to the identification of various behavioural and cognitive dimensions of teamwork, all of which are likely to have some sort of impact on DSA within a collaborative system. For example, there have been many attempts to postulate models of teamwork (e.g. Flieshman and Zaccaro, 1992; Helmreich and Foushee, 1993; Morgan et al., 1986; Salas et al., 2005 etc.). Most of the models presented attempt to define the different teamwork processes involved and also the different attributes that teams posses. Salas et al. (2005), for example, outlined the big five model of teamwork, arguing that the five most important teamwork processes are: leadership, mutual performance monitoring, back up behaviour, adaptability and team orientation. Salas et al. suggested that these factors would improve performance in any team, regardless of type, so long as three supporting mechanisms were also present within the team: shared mental models, closed loop communication and mutual trust. Morgan et al. (1986; cited in Salas et al., 1995) identified the following seven behavioural dimensions of teamwork: giving suggestions and criticisms, cooperation, communication, team spirit and morale, adaptability, coordination and acceptance of suggestions and criticisms. Salas, Burke and Canon-Bowers (2000) suggest that teamwork comprises the following processes:

- adaptability;
- shared situational awareness;
- oerformance monitoring and feedback;

- leadership and team management;
- interpersonal relations;
- coordination;
- communication; and
- decision making.

Salas et al. (1995) argue that team SA is interwoven with teamwork; however, despite this there has been little consideration of the impact of team processes on team SA (Salas et al., 1995). It is argued that the efficiency of teamwork attributes, behaviours and processes are all likely to have some impact on a system's level of DSA. Salas et al. (1995), for example, suggest that those team processes that facilitate communication (e.g. leadership, assertiveness and planning) will build SA. Inadequate communication between team members may result in shortfalls in the system's DSA. Inadequacies in any team processes are likely to adversely affect the system's DSA in some way. For example, the findings from the energy distribution case study demonstrated how lack of communication between agents in the network often led to the system's DSA being 'out of date' or behind the actual situation. Also, a lack of shared mental models between team members in the MNE4 experiment often meant that participants did not know who held the information that they required to update their SA, which often led to the system's DSA being becoming impoverished.

System Factors

Various aspects of a system's design can enhance or degrade a system's DSA. Endsley (1995a) pointed out the important role that system design has to play in the development and maintenance of SA, suggesting that a system either may not acquire all the required information, may fail to present the appropriate information to the operators who require it or that there may be incomplete or erroneous transmission of information to operators within the system. Based on the case studies conducted as part of this research, the system design factors that are likely to impact DSA include the interface design of the artefacts that are used to present SA-related information to the agents within the system, including the type of information presented, the manner in which the information is presented, who the information is presented to and how accurate the information that is presented actually is. For example, the findings derived from the land warfare digital mission support system studies indicated that there were inadequacies with the way in which the information was presented to users, who the information was presented to and also the accuracy of the information presented, all of which adversely affected DSA during the activities observed. The structure of the network of agents involved and the communications channels that are available to the different agents comprising the system are also likely to have an impact on the quality of the system's DSA. Communication links are one of the critical factors in the acquisition and maintenance of DSA and it is important that the appropriate communication links are present within a system and are maintained throughout task performance.

The importance of procedures in the development of a system's DSA has also been demonstrated. The communication strategies that teams adopt are therefore likely to play a significant role in DSA development and maintenance. Procedures that enforce the communication of critical DSA-related information, such as instructions, work progress and situational updates are particularly important. Stone and Posey (2008), for example, suggest that each member's awareness of the current situation could be significantly reduced if communication is not appropriate among members. One approach typically adopted by distributed teams is closed loop communication (Salas et al., 2001), which involves the initiation of communication by a sender, acknowledgement of receipt of the information by the receiver and then a follow up by the sender to check that the message was interpreted as intended (Salas et al., 2001). The use of such strategies within teams is critical to ensure that communications are completed accurately. Wilson et al. (2007) suggest that the use of closed-loop communication techniques is critical to ensure that information is clearly and concisely transmitted, received and correctly understood. In the military domain, for example, Wilson et al. (2007) report that a variety of friendly fire incidents have occurred due to inadequately executed closed loop communications.

The critical role of clear and appropriate procedures in the update and maintenance of DSA has also been demonstrated through this research. For example, the findings derived from the energy distribution case study indicated that procedures played a key role in the acquisition and maintenance of DSA. In this case the procedures dictated that the COCR operator communicated work instructions to agents in the field who then had to read back all the instructions received to confirm accurate receipt of them. In addition, the procedures dictated that the agents in the field undertaking the maintenance activities would periodically report to the COCR in order to give work and progress updates. The latter was particularly important in the maintenance of the system's DSA. Also, during the MNE4 case study it was found that unclear procedures (i.e. the CONOPS) led to confusion over how SA-related information was to be communicated around the system. In this case, the ambiguous nature of the CONOPs led to the different groups being confused over the exact nature of the KBD group's role. Some of the groups felt that it was the KBD group's role to proactively provide them with the information required for DSA, whereas the KBD group felt that their role was to wait for information requests. This confusion over procedures led to DSA being degraded during the early stages of the experiment.

Task Factors

The characteristics of the tasks being performed by teams can either facilitate or inhibit team performance (Paris et al., 2000). Various factors related to the tasks being performed are also likely to affect DSA. Factors such as task design, complexity, workload, time pressure, task allocation and familiarity with the task can all potentially affect the DSA acquired during performance of the task in question. For example, the level of workload experienced by team members is a key element in the safety, reliability and efficiency of complex sociotechnical systems (Gregoriades and Sutcliffe, 2008). Inappropriate levels of workload (either too high or too low) are likely to lead to reduced levels of DSA. Inappropriate levels of workload imposed on even one team member can affect the performance of the team as a whole (Roby and Lanzetta, 1957a, b and Dyer 1984;

all cited in Paris et al., 2000). However, the exact nature of the link between workload and SA remains ambiguous and so further exploration is required.

Endsley (1995a) points out that a major factor creating a challenge for operator SA is the increasing complexity of many systems and suggests that complexity can negatively affect SA via factors such as increased system components, the degree of interaction between components and the dynamics or rate of change of the components.

Summary

The purpose of this chapter was to present a model of DSA in collaborative systems. According to the model, DSA in collaborative systems can be viewed as the system's collective knowledge regarding a situation that comprises each agent's compatible awareness of that situation. The level of SA held by the components of the system is compatible, rather than shared, since it is developed based on distinct goals, roles, tasks, experience and schema. The knowledge required for DSA is acquired via the use of information that is held by systemic elements and passed around the system as and when required via SA transactions. According to the model, four key concepts underpin DSA, namely schema, the perceptual cycle, compatible SA and SA transactions.

Whilst the ideas presented in are quite different to those expressed by the dominant models of individual and team SA presented in the literature (e.g. Endsley, 1995a; Endsley and Jones, 1997) it is contended that that they are more appropriate for the study of SA in collaborative environments. The schema-based account of SA in collaborative systems affects existing models in four critical ways.

Firstly, using schema theory as a basis, it is argued that individual team members experience a situation in different ways and therefore that their awareness is compatible rather than shared. Each team member's SA is defined by their own personal experience, goals, roles, training, knowledge, skills and so on. 'The situation' can indeed be (objectively) defined in all manner of ways but under a schema/systems perspective there is a certain futility in this. Instead, SA is argued as being a systemic property (labelled the phenotype) which is the product rather than the sum of each individual's schema based 'theory of the world' (labelled the genotype).

Secondly, this account is in direct contradiction to those that suggest teams possess 'shared SA' (which tacitly assumes 'identical' awareness and an objectively definable situation). The DSA approach suggests that teams instead hold compatible and transactive SA. Within collaborative systems, each team member does not need to know everything; rather, they possess the SA that they need for their specific task. Yet they are also cognisant of what other team members need to and do know. Although different team members may be aware of the same information, this awareness is not shared, since the team members often have different goals and so view the situation differently based on their own task and goals. Each team member's SA is, however, compatible. This is the nub of DSA. It is different in content but is compatible in that it is collectively needed for the team to perform the collaborative task successfully. On the one hand it could be argued as to how all these individual heterogeneous experiences of the situation ever coalesce into something meaningful? The inconvenient truth, as it were, is simply that they do (as the case study discussed shows).

The idea of transactive SA is put forward, thirdly, as the means by which this occurs. Transactive SA focuses on transactions; elements and entities from one model of a situation can form an interacting part of another without any necessary requirement for parity of meaning or purpose. Thus transactions represent an exchange in awareness between team members. As stated above, it is the systemic 'transformation' of situational elements as they cross the system boundary from one team member to another that bestows upon team SA an emergent behaviour. The analytic and methodological challenge seems to be to ensure that this emergent behaviour is 'desirable'.

Fourthly, and finally, it is argued that there is significant utility in the progression from linear, feedback models of cognition (of the sort that underlies Endsley's three level model) in favour of a cyclical, parallel, generative model based on schema theory. This is a model that helps to explain why individuals can predict before they perceive (because they have pre-existing schemata), why less conscious reporting of SA probes can mean better SA (because schemas are often not available for conscious inspection and retrospective recall) and how individuals play a large part in creating better situations for themselves to be aware of (because the model is iterative and cyclical). The intuitive appeal of this approach is borne out in the case studies and further highlighted how deterministic models of SA to the probabilistic behaviour of teams.

The model also suggests that various factors affect the make up and quality of a system's DSA. These include individual factors (e.g. goals, roles, experience and training), team factors (e.g. level of teamwork, communication, collaboration, team competencies etc.), task-related factors (e.g. task complexity, workload etc.) and factors related to the system's design (e.g. support for SA requirements, communications links, technology used etc.).

Chapter 10
Conclusions for Distributed Situation Awareness Theory, Measurement and Teamwork

Introduction

The overall aim of this book was to explore and extend the concept of SA in collaborative environments. This involved using the foundations of a DSA theory laid by Stanton et al. (2006) to investigate the concept through a series of naturalistic case studies. A summary of the main conclusions derived from this overall programme of research is presented below, following which the implications of the findings are discussed.

What was Found?

Before going on to discuss the implications of this overall body of research, it is first worthwhile to summarise succinctly what was discovered. Initially, the reviews of SA models and SA measurement approaches served to highlight the significant level of contention surrounding the concept. The literature was found to be disparate, disjointed and divided and it was concluded that there are currently no universally accepted definitions or models of individual or team SA. In addition, existing SA theory was found to be inadequate for describing SA in collaborative environments and it was concluded that there are no suitable means available for measuring team SA during real world collaborative tasks. All of the SA models and measures presented in the literature were found to be inadequate for various reasons. Of most significance, perhaps, is the fact that the majority of the models and measures presented are individual operator-based (e.g. Bedny and Meister, 1999; Endsley, 1995a, b; Smith and Hancock, 1995); they focus exclusively on SA 'in-the-head' of individual operators either in terms of what it comprises, how it is attained or both. Whilst not problematic in itself (individual models and measures are wholly satisfactory when looking solely at individual SA), this becomes a significant problem when they are applied to collaborative systems. By its nature, teamwork is a complex phenomenon; it is 'a multidimensional, dynamic construct that refers to a set of interrelated cognitions, behaviours and attitudes that occur as team members perform a task that results in a coordinated and synchronised collective action' (Wilson et al., 2007, p. 245). Team SA therefore transcends the boundaries of individual operators and viewing SA only in the heads of individual operators is unsatisfactory, since it overlooks much of the workings of team SA in terms of the interactions between team members. SA arises from the interactions between operators and between operators and the technology that they use; it is associated with individual agents but it may not reside within them as it is born out of the interactions between them.

As a way forward, it was concluded that recently formulated DSA models were the most appropriate to drive research into SA in collaborative environments. We argued that systems based DSA models are more appropriate, since they consider the entire system, comprising the human operators and the tools that they use, and the interactions between them. Stanton et al. (2006) laid the basis for a model of DSA by outlining a series of theoretical foundations; however, this did not extend to prescribing a complete model of DSA for collaborative environments.

In order to drive the research forward, an extended model of DSA, based on Stanton et al.'s (2006) approach, and an accompanying modelling approach, the propositional network methodology, were presented and demonstrated. It was concluded that the model, along with its sub-concepts of compatible and transactive SA, required further validation and exploration through real world study. Further, it was concluded that the propositional network methodology required validation through applications in the real world.

A series of case studies was subsequently used to investigate the concept of DSA further and to test the propositional network DSA measurement approach. Each case study yielded significant findings in relation to DSA theory, but perhaps the most striking finding from each was that DSA ostensibly consisted of each team member's different but *compatible* portion of SA for the task in question. This finding is in direct contradiction to current team SA models (e.g. Endsley, 1995a; Endsley and Robertson, 2000), which suggest that team SA comprises team members' shared SA of the situation and that good team performance is dependent upon each team member's SA being the same for shared SA requirements. Further, the case studies undertaken suggested that, rather than sharing awareness, agents in collaborative systems engage in SA 'transactions' whereby SA-related information is exchanged between parties. The act of passing awareness onto another agent serves to modify the receiver's SA. Both parties are using the information for their own ends, integrated into their own schemata, and reaching an individual interpretation. Thus the transaction is an exchange rather than a sharing of awareness. The findings from the four case studies suggest that, on the contrary to shared SA accounts, good team performance is likely to be facilitated by supporting compatible SA requirements and SA transactions between team members through system and procedure design.

Based on the findings derived from this research, a model of DSA in complex collaborative environments was presented. The model is underpinned by four key concepts: schema theory, the perceptual cycle, compatible SA and transactive SA. It represents a cyclical, parallel, generative model of SA based on schema theory and postulates that DSA comprises each agent's compatible view of the situation and is built and maintained via SA transactions between agents. Team members each experience a situation in different ways (as defined by their own personal experience, goals, roles, training, knowledge, and skills and so on) and therefore their awareness is compatible rather than shared. In this view, SA is viewed as a systemic property which is the product rather than the sum of each individual's schema based 'theory of the world'.

These findings can be explored further using a number of key questions that were encountered throughout the conduct of this research.

Is the Distributed Situation Awareness Approach Useful for the Analysis and Design of Collaborative Systems?

This research has demonstrated that the DSA approach is suitable for describing and analysing SA during real world collaborative tasks. This is something that has thus far proved difficult for the HF community and the most prominent team SA models (e.g. Endsley and Robertson, 2000; Salas et al., 1995) and measures (e.g. Endsley, 1995b; Taylor, 1990) have been criticised for their inability to satisfactorily assess team SA (e.g. Artman and Garbis, 1998; Gorman et al., 2006; Patrick et al. 2006; Salmon et al., 2006, 2008a; Stanton et al., 2006; 2009; Shu and Furuta, 2005; Siemieniuch and Sinclair, 2006; Sonnenwald et al., 2004). In particular, the summation of individual team member SA in order to describe and assess team SA is problematic. Further, the most prominent team SA models and measures assess team SA typically only via SME interviews (e.g. Endsley and Robertson, 2000) or simulation and cannot be used to assess team SA during real world activities. Despite the complexities associated with teamwork, the DSA approach has shown itself to be capable of comprehensively describing and assessing DSA during real world collaborative activities. Its systemic viewpoint allows SA to be viewed in its entirety (rather than its component parts, as other models permit) as a non-linear emergent property of such systems that arises from the interactions between team members and the artefacts that they use. This in turn allows the collective information underlying DSA to be described, which in turn can be broken down further so that each agent's usage of, and contribution to, the information underlying DSA can be accounted for. Perhaps the main benefit of this is that only coordinated activity is considered, and therefore team and not individual agent SA is analysed.

The approach's utility lies in its outputs. Describing a system's awareness in this manner (i.e. information elements, their links with one another and their usage) not only allows the system's DSA to be described in terms of content (which can also be extrapolated to an individual level) and in terms of concepts and the relationships between them, but it also allows the description of differing views on the same situation by different agents. In this way, it goes much further than merely describing what pieces of information individuals need to know in order to perform tasks successfully. In particular, the mapping between information elements is a key output of the DSA approach. In addition, breaking a system's awareness down into information elements and the links between them allows judgements to be made on how well a system permits the communication, understanding and usage of the key information underlying DSA. This understanding of the key pieces of information underlying task performance and the links between them, who uses what information, in what manner and at what time throughout a scenario can potentially inform the design of more efficient systems, procedures and training programmes.

What are the Main Differences between the Distributed Situation Awareness Approach and the Shared Situation Awareness Approach?

The notion of compatible SA is in direct contradiction to the shared SA approaches advocated by others in the field (e.g. Endsley and Robertson, 2000). The DSA approach postulates that, within collaborative systems, team members have different, but compatible, SA regardless of whether the information that they have access to is the same or different. Shared SA accounts, on the other hand, suggest that some SA requirements are shared and that efficient team performance is dependent on team members having the same SA on shared SA requirements. Simply put, the DSA approach contends that not only is this not the case, but also that this may not be possible (in some cases). Furthermore, if it was the case then team performance may actually suffer rather than benefit.

The concept of SA transactions, how SA is exchanged between team members and how team member SA is modified as a result of these exchanges is also novel. This suggests that as team members receive information, its subsequent linkage with already held information leads to SA being modified. Thus, even when two team members have access to the same information, they use and view the information in a very different manner since it is the relationship between concepts that makes up their SA. Indeed, thinking about SA as the relationship between concepts is the key to the DSA approach; even when team members have access to the same information, the relationships between the information elements is likely to be different based on how they are using the information and what they need it for. It is this unique combination of information elements by each team member that makes their SA compatible and not shared. The very fact that an actor has received information, acted on it, combined it with other information and then passed it onto another actor means that its interpretation changes per team member.

Both approaches (DSA and shared SA) have their strengths and weaknesses. There can be little doubt that at least some proportion of the problem space can be tackled with a linear approach to SA and good results obtained (thus it is well worth the effort). This leaves the remainder of the problem space and it is here that the DSA approach comes to the fore. The main strengths of the DSA model are related to the systemic approach that it advocates. Firstly, the DSA approach takes the system itself as the unit of analysis rather than merely the individuals undertaking activity within it; it views SA as a non-linear, emergent property of collaborative systems. The systems thinking approach is one that has become accepted as an approach of considerable utility and is now prominent within HF circles. Indeed, many have articulated the utility of taking the overall systems as the unit of analysis rather than the individuals within the system (e.g. Hutchins, 1995; Hutchins and Klausen, 1996; Hollnagel, 2001; Hollnagel and Woods, 1999; Ottino, 2003 etc.). Further, any SA description needs surely to consider the technological as well as human agents residing within the system and the SA-related information that they bring to the table. Viewing SA in this manner:

- permits a systemic description of the information comprising SA (which can be extrapolated to an individual SA level);

- allows judgements to be made on potential barriers to SA acquisition and maintenance;
- enables team SA within complex collaborative systems to be viewed in its entirety, rather than as its component parts (i.e. individual and team member SA); and
- has the beneficial side effect that coordinated activity can be considered.

The DSA approach also has a strong theoretical underpinning, notably schema theory (e.g. Neisser, 1976), distributed cognition (Hutchins, 1995) and cognitive systems engineering (Hollnagel, 1998). One of the main criticisms of alternative SA models relates to their lack of theoretical underpinning. Endsley's model, for example, is based on the already contentious notions of information processing (Uhlarik and Comerford, 2002) and mental models (Smith and Hancock, 1995). In this case, the use of schema theory underlying the DSA concept gives the model a cyclical, parallel, generative nature which serves to explain why individuals can predict before they perceive (because they have pre-existing schemata) and how individuals play a large part in creating better situations for themselves to be aware of (because the model is iterative and cyclical). Finally, the DSA approach is also amenable to accurate assessment. By gathering verbal transcripts, task analyses, interview and cognitive task analysis data, it is possible to effectively determine what DSA comprises by way of identifying the underlying information elements and the relationships between them, what information was used by whom and what information was passed between different elements of the system. Taken collectively this provides a very powerful description of system endeavor.

The main weaknesses of the DSA approach are related to its complexity and its measurement. The approach is more complex than other team SA models; the departure in moving from thinking about individuals and what they know to thinking about the system and what it knows may be a difficult step to take. Further, since it is currently an emerging concept, much more investigation is required, although considerable evidence for the approach has been collected (e.g. Salmon et al., 2008; Stanton et al., 2006; Stewart et al., 2008; Walker et al., 2006). Questions may also be raised over the methodological aspects of measuring DSA. Firstly, unlike existing approaches such as SAGAT (Endsley, 1995b) and SART (Taylor, 1990), the propositional network approach does not quantitatively assess the quality of the system's and individual agent's SA. Therefore, judgements on the quality of DSA are made based on content analyses, task performance and SME and analyst subjective judgement. Secondly, the data used to identify the key information elements (e.g. verbal transcripts, CDM interview response data, observation transcripts etc.) can be criticised for its inability to identify the tacit SA-related knowledge (i.e. knowledge used but not openly expressed). However, the level of SME input reduces the potential for missing data in this case. Further, it could be argued that the relationships between the information elements (as depicted by the links within the propositional networks) provide a representation of this tacit knowledge. Finally, when CDM data is used, it is typically collected post-task performance and so could potentially suffer from the various problems associated with post-trial data collection, such as memory degradation (Klein and Armstrong, 2004).

Endsley's shared SA view, on the other hand, takes its main strengths from the simplicity of its approach. The approach suggests that in teams, some information requirements are distinct and some are shared or overlapping. On the face it, this view is correct; at a very high level of analysis, teamwork consists of both teamwork tasks (tasks where individuals interact or coordinate behaviour to undertake tasks important to the teams goals) and taskwork tasks (tasks being performed by individual team members in light of their individual roles within the team) and so it is logical to assume that, at a high level of description, some SA requirements will be the same across team members and that some will be distinct. This view, however, does not consider how the different team members are using the information and how their roles, tasks and experience affect their SA. The DSA approach contends that, in such cases, team member SA may be different even when they have access to the same information. Schema theory suggests that an individual's SA (regardless of whether the information used to build it is identical or entirely different) will be highly personalised based on experience, goals, roles, tasks, knowledge and schema. Of course, depending upon the environment under analysis, either approach may be correct; however, for complex modern day collaborative environments it is the author's view that more sophisticated approaches are required. Endsley's shared SA view also has an abundance of supporting research and has been applied in a wide variety of domains, including; aviation maintenance (Endsley and Robertson, 2000), the military (e.g. Endsley and Jones, 1997; Riley et al., 2006), aviation and air traffic control (Farley et al., 2000) and process control (Kaber and Endsley, 1998) to name only a few.

The main criticism of the shared SA approach concerns the concept of shared SA itself. According to Endsley and Robertson (2000), successful team performance requires that not only does each team member have good SA on his or her individual requirements, but also the same SA across shared SA requirements. We contend that, not only is this almost impossible (due to the reasons cited above), it is also typically not required within collaborative systems; teamwork relies on team members performing different activities and whilst they do need an appreciation of other team members' tasks and awareness, they do not necessarily want to develop the same SA as other team members. Further, we prefer to label shared SA requirements instead as transactive SA requirements in that the same information may be required by different team members, but it may often be used entirely differently. The very nature of team performance is such that different team members have different roles and so need to view and use information differently to other team members. As Gorman et al. (2006) point out, it does not make sense for everybody in a team to be aware of the same thing; rather, it is more important to ensure that the appropriate information is communicated to the appropriate team member at the right time.

Endsley's shared SA approach is also often criticised since it is based on her three level model account of individual SA and therefore does not consider team performance and the interactions between team members in any detail. In addition, many have pointed out that the three level model lacks a sound theoretical underpinning. For example, Smith and Hancock (1995) suggest that Endsley's reference to mental models, which themselves are ill-defined, is problematic and Uhlarik and Comerford (2002) criticised Endsley's theory for its use of an information-processing model containing psychological constructs that are not yet fully understood and that are subject to great debate themselves. The shared SA view also has weaknesses related to its accompanying measurement approach. It is difficult to apply SAGAT during real world collaborative tasks and it is difficult to generate appropriate SA probes in complex systems where SA requirements may not be accurately discernable prior to task performance.

What are the Implications of the Distributed Situation Awareness Approach?

The DSA approach has a number of significant implications for collaborative systems. Central to these are the notions that team SA comprises the compatible, rather than shared, SA of different team members and that team members engage in SA transactions. This means that it may be more appropriate to design collaborative systems that cater for compatible, rather than shared, SA and to provide systems that support SA transactions between different team members. Currently, many collaborative systems are designed so that all the information required by the team is available to every team member (such as the mission support system analysed in Chapters 7 and 8); indeed proponents of NEC systems project that enhanced levels of information sharing will lead to enhanced levels of SA. The DSA approach suggests that this is inappropriate, and that DSA will be enhanced more by providing interfaces and displays that present only the SA information that is required by each team member. The DSA approach advocates a 'right information at the right time to the right team member' design philosophy.

Our approach suggests that the most prominent displays and interfaces should present only the SA information that is required by each user within the collaborative system. Users should be able to easily access the SA-related information required for their role and should not be inundated with redundant information (required by other agents but not themselves). Whilst this means that more sophisticated systems may be required (i.e. that can be customised or tailored based on the user using them) the findings from this research suggest that it will provide support that is more effective for DSA.

Further, SA transactions should be supported where possible. This means that designers need to know exactly what it is that different users need to know and what they need to know it for. To support SA transactions and DSA development, systems should present incoming information in conjunction with the information with which it is likely to be used. For example, a land warfare mission support system (analogous to the one analysed in Chapters 7 and 8) could present new incoming information regarding a destroyed combat vehicle to combat service support staff (whose job it is to remove and deal with casualties, repair damage and replenish forces) in conjunction with information relating to routes to and from the vehicle, casualty evacuation routes, distances and projected times, combat effectiveness, medical support information (e.g. nearest hospitals etc.), force replenishment requirements and also resource availability. In this way, the system is supporting the integration of the information from the SA transaction with the combat service support staff's existing awareness and future awareness needs. The same system could present the information regarding the destroyed combat vehicle very differently in the light of different user needs. For example, when presenting the information to the Chief of Staff (who is 'running' the battle at the ops table), associated information presented could include the proximal units and their capabilities, the commander's effects schematic, the task ORG and the combat service support staff's assessment. This information would then support the Chief of Staff in allocating the destroyed unit's tasks to another unit on the battlefield.

The DSA approach therefore suggests that it is critical that the system design process begins with a clear definition of the compatible and transactive SA requirements of the different components of the collaborative system. These can be identified using approaches such as HTA and the CDM. The SA requirements specification should then be used to drive the design of distinct systems (i.e. displays, interfaces and tools) for each element of the collaborative system. Only through this process can DSA be truly supported by system design. Without such an approach, users may be overloaded with redundant information and tools.

This conclusion is corroborated by other similar recommendations presented in the literature. In a similar case study, for example, Bolstad, Riley, Jones and Endsley (2002) found explicit differences between the SA requirements of US Army Brigade officers. In conclusion, they recommended that displays should be tailored to each officer's needs whilst also providing information relating to the SA of the other officers in the team. Along the same lines, Gorman et al. (2006) suggest that it may in fact be prohibitive and counteractive to give everyone mutual access to the same information. In conclusion to an analysis of the Gulf War Black Hawk friendly fire incident, Gorman et al. (2006) discussed the typical team SA design principle that every team member should be presented with all the information that is relevant to the team as a whole. In conclusion, they reported that:

> this design principle breaks down in command and control environments as the size of the team increases and as team members have more specialised roles, where it may be prohibitive and counteractive, respectively, to give everyone mutual access to the same information. (pp. 1322–3)

Similarly, Kuper and Giuerelli (2007) postulate that in order to enhance command and control team efficiency, tailored work aids should be used to reduce the cognitive load associated with mining through redundant information. They argue that the key to efficient and effective command and control team performance is the design of work aids that support both holistic work practices and unique first person perspectives.

The findings also suggest that other means can be taken to enhance DSA. These include the use of well thought out and enforced procedures (e.g. dictating that key information is communicated to the key agents involved), clearly defined roles and responsibilities (e.g. that make explicit who possesses what information and how it can be accessed) and the presence of appropriate communications links can all be used to enhance DSA in collaborative systems.

Measuring Situation Awareness in Collaborative Systems

One of the main aims of this work was to develop and validate a suitable approach for both describing the content of and assessing DSA during real world collaborative activities. The propositional network approach was put forward as an approach that could satisfy both requirements. The findings derived from this research, particularly the case studies, have led to a number of conclusions regarding the propositional network approach. In its present format the approach has demonstrated that it can be used to accurately describe a system's DSA and the usage of the information underlying DSA by the different agents involved. Of particular novelty is the way in which the propositional networks describe not only the information elements underlying DSA, but also the mapping between the different information elements; this is something that so far has not been supported by existing SA measures. Further, the key information underlying SA can also be identified using social network metrics or the five plus links rule. The approach avoids most of the flaws that are typically associated with the measurement of SA (see Chapter 3). Since the data is obtained via observation or post-task interview, the task under analysis is not affected in anyway (i.e. no freezes of the task are required). In addition, since no probes are used there is no requirement to develop appropriate probes *a priori*. Further, since there are no subjective ratings of the quality of SA the approach is not beset by the flaws typically associated with the subjective rating of SA, such as correlations with performance, memory degradation and lack of awareness of low awareness portions of the task. Finally, the validation of the propositional networks via SMEs removes any doubts regarding the accuracy of the DSA description.

In addition, this research has demonstrated how propositional networks can be used to assess and inform system and interface design. Systems can be assessed to the extent to which they present each of the information elements contained in the propositional networks, whether they support the mapping between SA information elements and also whether they present the information in an accurate and timely manner. This provides a useful framework for system and interface design assessments. In addition, the links between information elements contained within propositional networks can be used to determine what information should and should not be grouped together on interfaces.

This research has also led to insights into ways in which the propositional networks approach can be improved for future DSA assessment. Firstly, although it was originally suggested that propositional networks should be developed from CDM interview response data, constraints imposed during the case studies conducted (i.e. limited time and access to SMEs) led to different avenues being pursued in order to collect the data required. Subsequent SME reviews of the propositional networks developed suggest that they did not suffer in any way due to the use of different data inputs. It is therefore recommended that propositional networks can also be developed from other data sources, including verbal transcripts, HTA descriptions and SOIs. Further, propositional networks can also be supplemented by observational transcript data. Secondly, propositional networks can be meaningfully analysed using social network metrics in order to identify key information elements within a system. This information is useful as it can be used to inform system design, i.e. ensuring that the most important information is prominent on displays and interfaces. Thirdly, and finally, this research has indicated that, when propositional networks are large, complex and unwieldy, summary propositional networks containing high level information elements are useful as representations for communicating analysis findings.

Distributed Situation Awareness and System Design

Endsley (2004) suggests that, 'the most interesting frontier for SA remains in the design arena' (Endsley, 2004, p. 337). One of the key aims of this research, and indeed one of the major challenges for the concept, therefore, is to transform what we know about SA into guidance for how to design systems so that they enhance, rather than inhibit, the SA of teams working within them. To this end, a DSA design process and design guidelines for DSA, developed based on the findings derived from this research, are presented below.

Distributed Situation Awareness Design Process

The findings derived from the research undertaken can be used to formulate an overall process for system designers wishing to develop systems (e.g. mission support systems) that support DSA acquisition and maintenance during collaborative activities. A flowchart depicting a DSA design process, derived from a synthesis of the findings of this research, is presented in Figure 10.1.

The process begins with the conduct of an SA requirements analysis in order to comprehensively identify and record all the SA requirements of the different end users of the system in question. This involves the conduct of a HTA (Annett et al., 1971) of the system in question, using data derived from observations of the existing system, SOIs, interviews with SMEs, training manuals and other appropriate documentation. The HTA break down should then be used to identify the system's SA requirements. Following this, propositional networks should be constructed in order to identify the relationships between the different SA requirements. It is important to note that the SA requirements analysis phase does not involve merely identifying the different pieces of information that need to be known; rather, it involves going further and identifying what it is that needs to be known, how this information is used and what the relationships between the different pieces of information actually are, i.e. how they are integrated and used by different users. In particular, identifying the relationships between different pieces of DSA-related information allows designers to group information meaningfully in their end design.

Following the SA requirements analysis phase it is next important to identify which of the information elements underlying DSA represent compatible SA information elements (i.e. used in a different way by different team members), which are transactive SA information elements (i.e. passed between team members during the process in question) and which are both. This involves taking the SA requirements analysis outputs (i.e. HTA and propositional networks) and, in conjunction with SMEs for the process in question, classifying each information element accordingly.

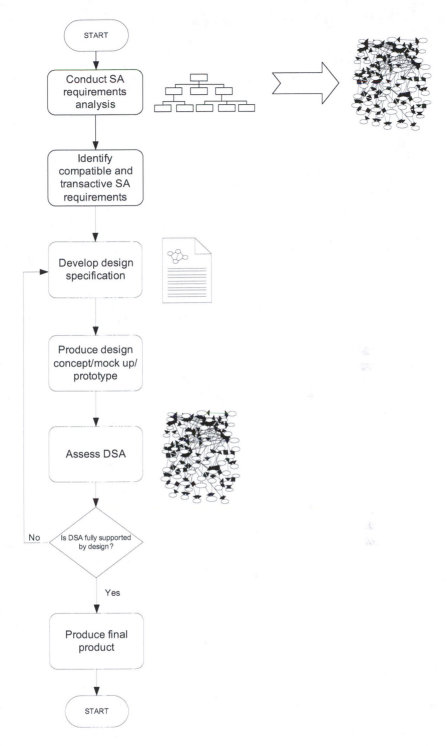

Figure 10.1 Distributed situation awareness design process flowchart

The SA requirements analysis outputs and the compatible and transactive SA elements classification should then be used to inform the development of an SA oriented design specification. Again, it is important here to note that this should not involve merely specifying what it is that the system should be presenting to its different users. Rather, this should involve a specification of what it is the system should be presenting to whom, in what format the information should be presented, what other information it should be presented in conjunction with (i.e. the relationships between different classes of information) and also what the information is to be used for (which may include many different things in collaborative systems). To support DSA, it is important that the SA design specification informs designers about the processes that the SA-related information is being presented to support and about the other information that information presented is likely to be used in conjunction with.

Based on the design specification, mock ups should then be developed. Following this, DSA testing should begin. It is critical that SA is tested throughout the design lifecycle if SA requirements are to be supported by the end design. Without such testing, it is impossible to determine whether the design supports or hinders DSA acquisition and maintenance. The DSA assessments should be undertaken using the propositional network methodology. The assessment should involve determining the extent to which the design supports the DSA requirements specified, although the exact nature of the assessment is dependent on the stage that the design has reached. For example, at the mock up stage, the assessment might entail walking through the task with SMEs and evaluating the extent to which SA requirements, and the relationships between SA-related information elements, are supported by the interface in question. On the other hand, when the design is at a prototype stage, the assessment could involve assessing user performance and DSA during actual operational trials. In addition to the propositional network approach, this might also involve conducting interviews (i.e. CDM interviews) with SMEs post-trial.

Any DSA-related deficiencies found during DSA testing then send the design back into the iterative phase of developing specifications, producing concepts and mock ups and then testing them. This cycle also continues through mock up, concept and prototype versions of the system in question. Only when designers are satisfied that DSA requirements are fully supported can the prototype design proceed from this cycle into the development of the final product.

Distributed Situation Awareness Design Guidelines

Endsley et al. (2003) correctly point out that traditional HF design guidelines are inadequate for achieving the SA required in complex systems since they typically address the physical and perceptual characteristics of systems rather than the way in which systems should function from a cognitive standpoint. Further, there are only limited SA-specific design guidelines available (e.g. Endsley et al., 2003). In response to this, the following core collaborative system design principles can be extracted from the findings derived from our research into DSA (it is worth pointing out here that the guidelines presented focus on enhancing DSA rather than enhancing SA 'in-the-heads' of individual operators).

1. *Clearly define and specify SA requirements.* The importance of system designers knowing what it is that different users of systems 'need to know' during task performance has been demonstrated throughout this research. The collaborative system design process should therefore begin with a clear definition and specification of the DSA requirements of the overall system and of the different operators working in the system in question. This should include a breakdown of both compatible and transactive SA requirements. As stated above, it is important that DSA requirements analysis specification includes more than just the different pieces of information that need to be known and should go further to describe what it is that needs to be known and by whom, how this information is used by different users and what the relationships between the different pieces of information actually are. i.e. how they are integrated and used by different users. Matthews et al. (2004) point out that knowing what the SA requirements are for a given domain provides engineers and technology developers with a basis on which to develop optimal system designs to maximize human performance rather than overloading workers and degrading their performance. It is recommended that DSA requirements analysis be conducted using the HTA and propositional network procedure described above, and that it should involve SMEs for the domain in question.

2. *Ensure roles and responsibilities are clearly defined.* Collaborative systems utilise teams consisting of multiple members, each of which have distinct goals, roles and responsibilities. The different roles and responsibilities within teams need to be clearly defined; each team member needs to fully understand what it is that they need to do and also be cognisant of what it is the other team members do. Whilst this important in terms of overall role and contribution to the team, it is also important on a minute-by-minute basis. This should include knowledge of what other team members are doing (tasks) but also knowledge of what other team members should know at that point in time (meta SA). Ambiguity in role definition can adversely affect DSA since it leads to confusion over who knows what at what times and who possesses what information. The MNE4 case study, for example, demonstrated how DSA can be adversely affected when roles are not clearly defined in collaborative systems.

3. *Design to support compatible SA requirements.* This research has demonstrated that team members working in collaborative systems each have distinct, but compatible, SA requirements. Collaborative systems should therefore be designed to cater for these compatible SA requirements, rather than to support shared SA between team members. Rather than present everything to everyone, or use common operational picture displays, this research suggests that collaborative systems should be designed so that users are not presented with information, tools and functionality that they do not explicitly require. Systems should therefore be designed to support the roles, goals and SA requirements of each of the different users involved in the process in question. This might involve the provision of different displays, tools and functions for the different roles and tasks involved, or might involve the use of customisable interfaces and displays.

4. *Design to support SA transactions.* This research has proposed the concept of SA transactions between agents as the means by which DSA is developed and maintained during collaborative tasks. Transactions in SA between team members involve the exchange of SA-related information elements and the subsequent integration of this information with existing schema. Systems and interfaces that present information to team members should therefore be designed so that they support SA transactions where possible. This might involve presenting incoming SA transaction information in conjunction with other relevant information (i.e. information that the incoming information is related to and is to be combined with) and also providing users with clear and efficient communications links with other team members. Similarly, procedures can be used to support SA transactions; this might involve incorporating certain pieces of information into procedural communications between team members in order to support SA transactions.

5. *Group information based on links between information elements in DSA requirements analysis.* The propositional networks produced as a result of the DSA requirements analysis highlight which elements of information are used together, at different points in the task, for DSA acquisition and maintenance. Designers should group information (on interfaces and displays) based on the links between information elements specified in the propositional networks that are produced as a result of DSA requirements analyses.

6. *Support meta SA through training, procedures and displays.* Meta SA – awareness of what other agents in the system know – has been found to be important throughout this research since it enables SA transactions to occur at the appropriate time between the appropriate team members. Meta SA should therefore be supported through the use of training, procedure and display design. Our research has highlighted the importance of team members having an understanding of what it is that other team members are doing and therefore should 'know' at different times during task performance; this allows team members to understand when and where information is required in the distributed team. It is therefore recommended that, through team training and system design interventions, each team member has an appreciation of what it is that the other team members need to know at which points during task performance. Stanton et al. (2006), for example, point out that 'it is important for the agents within a system to have awareness of who is likely to hold specific views and, consequently, to interpret the potential usefulness of information that can be passed through the network in terms of these views' (p. 1308). Further, Stanton et al. (2006) point out that there are two aspects of SA at any given node in a distributed team; individual SA of one's own task and 'meta' SA of the entire system's DSA.

7. *Remove unwanted information.* The case studies focusing on NEC and digital mission planning systems (Chapters 7 and 8) highlighted the problems associated with users having to plough through redundant information in order to locate the information that they required for DSA. Information that is not needed by team members should therefore not be presented to them; this includes menu structures

that are not required, tools that are not required and display information that is not required. Any displays or interfaces should therefore be designed so that, based on user requirements, unwanted information can be removed or hidden.

8. *Use customisable/tailored interfaces.* As articulated previously, the nature of collaborative systems is such that there are specific roles and SA requirements. Subsequently, the information and the tools that one agent needs to use may be very different to that needed by another agent. Collaborative system interfaces and displays should therefore be customisable, allowing users to customise the interface (either themselves or intelligently by the system based on usage) so that only the information and tools that they require are present. This increases the usability and ease of use of the system and reduces interaction time (i.e. having to mine through menus to find information and tools required).

9. *Use multiple interlinked systems for multiple roles and goals.* The analysis of the digital mission support system presented in Chapters 7 and 8 suggested that role specific systems might be more appropriate to support DSA development and maintenance. When a team is divided into distinct roles, team members have very different goals and informational requirements; it may therefore be pertinent to offer separate (but linked) support systems. In the same way that Microsoft Office provides separate word processing (e.g. Word), drawing (e.g. Visio) and spreadsheet (e.g. Excel) tools, distributed team working support systems should provide a suite of mission support tools catering for the different users and roles involved; each tool should have the functionality and information required for the role it is designed to support whilst also containing the ability to see global information.

10. *Consider the technological capability available and its impact on SA.* The analysis of the digital mission support system presented in Chapters 7 and 8 highlighted the problems associated with technology limitations that can degrade DSA (e.g. bandwidth limitations that led to DSA information presentation being untimely and inaccurate, and mapping and screen resolution issues that limit user SA of the battlefield). Again, perhaps an obvious, but nevertheless critical, recommendation is that system designers need to carefully consider the constraints imposed on them by technological capability and design the system accordingly within these constraints.

11. *Ensure that the information presented to users is accurate at all times.* Each of the case studies conducted as part of this research highlighted the importance of the communication of only accurate information. The transmission of erroneous information leads to erroneous DSA and a reduction in tempo, since measures are often taken to interrogate questionable information (as was the case during the battle execution activities analysed in Chapter 8). Efficient levels of DSA are ultimately contingent on accurate information. The information presented by any collaborative system should therefore be highly accurate and system designers need to ensure that the information presented by all aspects of the system is accurate at all times.

12. *Ensure information is presented to users in a timely fashion and that the timeliness of key information is represented.* The temporal nature of DSA and

the subsequent importance of keeping DSA up to date was emphasised by the findings derived from the studies presented in Chapters 5, 7 and 8. Chapter 8 in particular highlighted how information that is not presented in a timely manner can lead to DSA decrements. SA-related information should therefore be presented to users in a timely manner, without any delay, at all times. Further, the timeliness of information should be represented on interfaces and displays, allowing users to determine the latency of information.

13. *Provide appropriate and explicit communications links.* Communication is defined as 'the process by which information is clearly and accurately exchanged between two or more team members in the prescribed manner and with proper terminology; the ability to clarify or acknowledge receipt of information' (Cannon-Bowers et al., 1995, p. 345). Throughout this research, the importance of efficient, appropriate communications links as an enabler for distributed team working and DSA has been highlighted. It is critical that collaborative systems possess the appropriate communications links and that the users working with the system understand which communications channels are and are not open to them. This follows on from Stanton et al.'s (2006) conclusion that the links between agents in a network are more crucial than the agents themselves in maintaining DSA.

14. *More information is not always better.* Designers should reconsider the notion that presenting all the information contained within a system will lead to enhanced levels of DSA. Each of the case studies conducted suggests that it is preferable to present users with only the information that they specifically require rather than all of the information that the system is capable of presenting. Bolia et al. (2007) suggest that the increased amount of information available does not necessarily mean that users of the data will make better decisions due to a number of factors, including that increases in quantity of information do not necessarily lead to an increase in the amount of relevant information, the fact that all data is not good data and that false data could be deliberately be fed into networks or data could be erroneous and also that data is only as good as their interpretation.

15. *Use filtering functions.* When systems have common operational displays containing movement and location information relating to distinct entities (e.g. enemy, friendly, neutral etc.) on a map, it is important that the system allows the users to filter the display so that different classes of information only are displayed. This provision is also likely to support DSA development since the user can select the different pieces of information that they want the system to present together.

16. *Present SA-related information in an appropriate format.* The systems in which the DSA approach is typically applied are complex, dynamic and information rich and the human elements of these systems need to assimilate and understand large volumes of information. It is therefore critical that systems are designed to that the information presented to users is in a format that is amenable to quick, efficient and accurate assimilation and understanding. The information presented should be in a format that facilitates DSA development. Endsley (2000) points

out that designers must ensure that systems provide not only the information required by users, but also that the information is presented in a manner that is cognitively useable.

17. *Use procedures to facilitate DSA.* This research has indicated that procedures are an effective means of facilitating DSA acquisition and maintenance through SA transactions. It is therefore recommended that procedures should be used to support SA transactions via encouraging the continual communication of DSA-related information around collaborative systems and also by structuring communications so that related information is communicated together.

18. *Test DSA throughout the system lifecycle.* It is clear that DSA should be considered and tested where possible throughout the design lifecycle. Despite this, from our experiences this is not always the case; HF analyses such as SA assessments are often only undertaken as an afterthought once systems are operational and suggested redesigns may be too expensive to implement. It is therefore recommended that DSA is tested throughout the system design lifecycle, from the concept phase, through mock up and prototype phases to the operational end product. This might involve simply assessing DSA using functional diagrams of an interface and SMEs, or might involve large-scale operational trials of prototypical systems. The extent to which the system being designed supports the DSA requirements of its users should be evaluated throughout the entire design lifecyle.

Areas for Future Research

Throughout the course of this research, the following key areas for future research within the SA area emerged.

With any new model of human performance, it is important to develop a library of case studies; each new application can be used to validate the DSA approach and also to provide additional insight into the capabilities and limitations of the approach. In addition to the three domains focused on in this research, others in the field have used this approach to assess DSA in the naval domain (Stewart et al., 2008), air traffic control (Walker et al., 2005), rail (Walker et al., 2006), road transport (Walker et al., In Press) and the emergency services (Houghton et al., 2006). Within the realm of collaborative activity, this represents only a minute portion of the different domains present and it is therefore recommended that further DSA evaluations be undertaken across additional collaborative domains. Potential domains of inquiry include the nuclear power domain, civil aviation, team sports (e.g. football, rugby), the maritime domain and the gas and oil production domains.

Research undertaken to test systems and devices designed based on DSA principles is also recommended. The ultimate aim of developing theory and measures, within HF circles at least, is to provide knowledge that enables systems and artifacts to be designed better so that human performance is enhanced and not inhibited. With the implications for system design outlined above, one pertinent avenue of further exploration is to develop and test compatible or DSA-based interfaces and devices. In particular, 'compatible SA' oriented designs should be tested against 'shared SA' oriented designs in order to evaluate which affords the higher levels of SA in collaborative systems.

Although this and other research (e.g. Endsley et al., 2003) has attempted to prescribe guidance for collaborative systems designers, it is apparent that further, more specific guidance on how to design systems, interfaces and artefacts to enhance DSA is required. At present, the guidance presented here and in the literature is at a relatively high level, and further specific advice is required. It is therefore recommended that further investigation be used to formulate a series of specific and exhaustive DSA-oriented system design guidelines.

As with other HF constructs, the final frontier for SA is the ability to accurately predict it. The need for prescriptive SA models has been discussed by many in the area (e.g. Bryant et al., 2004; Endsley, 2004; Moray, 2004; Rousseau et al., 2004). The ability to predict, *a priori*, the level of SA afforded by a particular system or device design is an extremely powerful commodity. However, the provision of such models is likely to be extremely difficult and thus much more investigation is required. Endsley (2004) points out that 'while a worthwhile goal … it is certainly a tall order for SA' (p. 329). It is therefore recommended that further research is expended in exploring the prediction of DSA in collaborative systems using the propositional network approach.

The propositional network approach used throughout this research is still relatively new and although it served its purpose here, it is recommended that further study is required, not only to validate the approach further but also in methodological terms to investigate new means of analysing the networks developed. It is our opinion that the propositional network approach could potentially be extended in three areas. Firstly, one of the criticisms of the approach is that they do not provide a quantitative analysis of the quality of a system's DSA. Other SA measurement approaches tend to present some sort of quantitative analysis (e.g. SAGAT, SART). It is therefore recommended that investigation be made into how the propositional network approach can be modified so that DSA is quantitatively assessed. Secondly, it is our opinion that the data collection procedure used for propositional networks could be refined somewhat. Currently, input data is derived from observational study, verbal transcripts, HTA and/or standard operating instructions (either in combination or in isolation). It is our opinion that a set of specific probes designed to collect the data required for propositional networks could be developed, based on the experiences encountered during this research. Examples of the types of probes that could be used include:

- What were the goals of the different agents involved throughout the scenario?
- What activities did you (and the other agents involved) need to perform in order to achieve the scenario goals?
- Describe the information/knowledge that each of the agents involved 'needed to know' in order to accomplish their required tasks successfully. Where did this information/knowledge come from?
- What plans, strategies, procedures (e.g. Standard Operating Procedures) and work instructions were used throughout the scenario?
- What documents (e.g. Standard Operating Procedures, instructions, diaries, databases etc.) were used by different agents throughout the scenario?

- What equipment (e.g. tools, technology, displays, controls, databases) was used, by yourself and by the other agents involved, during the scenario?
- What specific locations were involved in the scenario (e.g. where were the agents located and where did the work activities take place)?
- Which agents were involved in the scenario?
- Of the agents involved, who did communications take place between throughout the scenario? What technology was used to mediate these communications? What was the content of these communications?
- What assets/resources did each of the agents involved have at their disposal throughout the scenario?

Thirdly, and finally, further investigation is required into the use of additional metrics to analyse the propositional network outputs. At the moment, social network metrics such as centrality and sociometric status are used to identify key information elements. It is our opinion that the use of further social network metrics should be investigated. For example, the network robustness metric refers to the degree to which a network can continue to function efficiently when one of its nodes is removed. This metric could be used to identify the impact on DSA when certain pieces of information are removed from a system.

A DSA-based design process was presented based on this research. The next logical step is to provide software support for the DSA-design process. It is therefore recommended that a DSA-design software tool be developed. Such a tool could provide support for DSA requirements analysis, DSA design (e.g. provision of DSA design guidelines) and DSA analysis (e.g. propositional networks).

Further research is also recommended regarding the effect of teamwork processes on DSA levels. Team performance is well defined in terms of the different processes underlying it (e.g. Salas et al., 2005); however, the literature reviews conducted as part of this research suggest that there has been only very limited investigation into the specific effects of each of these processes on team SA. For example, exactly what effect on team SA does inadequate coordination, communication, mutual monitoring and team leadership have? Whilst many have alluded to the likely effects, (e.g. Salas et al., 1995) it is apparent that there has been little scientific investigation of these effects. It is therefore recommended that studies focusing on the impact that inadequate teamwork processes have on SA during collaborative activity.

Closing Remarks

SA is a critical commodity for teams working within complex systems. With ever-increasing technological capabilities, there is great potential for enhancing DSA in collaborative systems through the provision of advanced technological support systems; however, this is accompanied also by a very real opportunity to create inappropriately designed systems that can hinder rather than support DSA development and maintenance. Enhancing SA in any complex sociotechnical system requires that we understand how SA operates in the environment in question and also what SA might comprise. In this book, we have presented a distributed cognition-based systems model of DSA and an accompanying modeling approach for complex sociotechnical systems and have argued that the both are more suited than existing models and measures to studying SA in such environments and informing collaborative system and procedure design. It is our hope that in doing so, this book has provided a significant contribution to knowledge in the area, in terms of how team SA should be viewed, how it can be assessed and what measures can be taken to enhance it. We hope that the model and method advocated by this research are taken, applied and advanced further by other researchers and that ultimately collaborative systems benefit from the resultant knowledge generated.

References

Adams, M.J., Tenney, Y.J. and Pew, R.W. (1995). Situation awareness and the cognitive management of complex systems. *Human Factors*, 37:1, pp. 85–104.

Alberts, D.S. and Hayes, R.E. (2006). Understanding command and control. Command and Control Research Program (CCRP), Washington DC.

Alberts, D.S. and Hayes, R.E. (2007). Planning: Complex endeavors. Command and Control Research Program (CCRP). Washington DC.

Alberts, D.S., Garstka, J.J. and Stein F.P. (1999). Network centric warfare. Command and Control Reseach Program (CCRP), Washington DC.

Anderson, R.C. (1977). The notion of schemata and the educational enterprise. In R.C. Anderson, R.J. Spiro and W.E. Monatague (eds), *Schooling and the Acquisition of Knowledge* (pp. 415–31). Hillsdale, NJ: Lawrence Erlbaum.

Anderson, J. (1983). *The Architecture of Cognition*. Cambridge, MA: Harvard University Press.

Annett, J. (2002). A note on the reliability and validity of ergonomics methods. *Theoretical Issues in Ergonomics Science*, 3:2, pp. 228–32.

Annett, J., Duncan, K.D., Stammers, R.B. and Gray, M.J. (1971). Task analysis. Department of Employment Training Information Paper 6. HMSO, London.

Artman, H. (2000). Team situation assessment and information distribution. *Ergonomics*, 43:8, pp. 1076–95.

Artman, H. and Garbis, C. (1998). Situation awareness as distributed cognition. In T. Green, L. Bannon, C. Warren and J., Buckley (eds), *Cognition and cooperation. Proceedings of Ninth Conference of Cognitive Ergonomics* (pp. 151–6). Limerick: Ireland.

Artman, H. and Wærn, Y. (1999). Distributed cognition in an emergency co-ordination center. *Cognition, Technology and Work*, 1:4, pp. 237–46.

Baber, C. (2003). *Cognition and Tool Use*. London: Taylor and Francis.

Baber, C. and Stanton, N.A. (1996). Human error identification techniques applied to public technology: Predictions compared with observed use. *Applied Ergonomics*, 27: 2, pp. 119–31.

Baber, C. and Stanton. N.A. (2002). Task analysis for error identification: Theory, method and validation. *Theoretical Issues in Ergonomics Science*, 3:2, pp. 212–27.

Banbury, S.P., Croft, D.G., Macken, W.J. and Jones, D.M. (2004). A cognitive streaming account of situation awareness. In S. Banbury and S. Tremblay (eds), *A Cognitive Approach to Situation Awareness: Theory and Application* (pp. 117–37). Aldershot: Ashgate Publishing.

Bartlett, F.C. (1932). *Remembering: A Study in Experimental and Social Psychology*. Cambridge: Cambridge University Press.

Bedny, G. and Meister, D. (1999). Theory of activity and situation awareness, *International Journal of Cognitive Ergonomics*, 3:1, pp. 63–72.

Bell, H.H. and Lyon, D.R. (2000). Using observer ratings to assess situation awareness. In M.R. Endsley and D.J. Garland (eds), *Situation Awareness Analysis and Measurement*, Mahwah, NJ: Laurence Erlbaum Associates.

Billings, C.E. (1995). Situation awareness measurement and analysis: A commentary. Proceedings of the International Conference on Experimental Analysis and Measurement of Situation Awareness, Embry-Riddle Aeronautical University Press, FL.

Blandford, A. and Wong, W. (2004). Situation awareness in emergency medical dispatch. *International Journal of Human–Computer Studies*, 61:4, pp. 421–52.

Bolia, R.S. (2005). Intelligent decision support systems in network-centric military operations. In Intelligent decisions? Intelligent support? Pre-proceedings for the International Workshop on Intelligent Decision Support Systems: Retrospect and prospects, pp. 3–7.

Bolia, R.S., Vidulich, M.A., Nelson, W.T. and Cook, M.J. (2007). A history lesson in the use of technology to support military decision-making and command and control. In M.J. Cook, J.M. Noyes and Y. Masakowski (eds) *Decision Making in Complex Systems* (pp. 191–200), Aldershot: Ashgate Publishing.

Bolstad, C.A. and Endsley, M.R. (2000). The effect of task load and shared displays on team situation awareness. In Proceedings of the 14th Triennial Congress of the International Ergonomics Association and the 44th Annual Meeting of the Human Factors and Ergonomics Society, 2000.

Bolstad, C.A. and Endsley, M.A. (2007). Measuring shared and team situation awareness in the army's future objective force. In the Human Factors and Ergonomics Society Annual Meeting Proceedings, Baltimore, pp. 369–73.

Bolstad, C.A., Cuevas, H.M., Gonzalez, C. and Schneider, M. (2005). Modeling shared situation awareness. Paper presented at the 14th Conference on Behaviour Representation in Modeling and Simulation (BRIMS), Los Angeles, CA.

Bolstad, C.A., Riley, J.M., Jones, D.G. and Endsley, M.R. (2002). Using goal directed task analysis with Army brigade officer teams. In Proceedings of the 46th Annual Meeting of the Human Factors and Ergonomics Society, Baltimore, MD, pp. 472–6.

Brehmer, B. (2007). Understanding the functions of C2 is the key to progress. *The International C2 Journal*, 1:1, pp. 211–32.

Brown, I.D. (2001). A review of the 'Looked-But-Failed-To-See' accident causation factor. Department of Environment, Transport, and the Regions Conference on Driver Behaviour, University of Manchester.

Bryant, D.J., Lichacz, F.M.J., Hollands, J.G. and Baranski, J.V. (2004). Modeling situation awareness in organisational context. In S. Banbury and S. Tremblay (eds), *A Cognitive Approach to Situation Awareness: Theory and Application* (pp. 104–16), Aldershot: Ashgate Publishing.

Burke, S. C. (2004). Team task analysis. In N.A. Stanton, A. Hedge, K. Brookhuis, E. Salas and H. Hendrick (eds), *Handbook of Human Factors and Ergonomics Methods* (pp. 56.1–56.8). CRC Press: Boca Raton, FL.

Cannon-Bowers, J.A. and Salas, E. (1997). Teamwork competencies: the interaction of team member knowledge skills and attitudes. In O.F. O'Neil (ed.), *Workforce Readiness: Competencies and Assessment* (pp. 151–74). Hillsdale, NJ: Erlbaum.

Cannon-Bowers, J.A., Tannenbaum, S.I., Salas, E. and Volpe, C.E. (1995). Defining competencies and establishing team training requirements. In R.A. Guzzo and E. Salas (eds), *Team Effectiveness and Decision making in Organizations* (pp. 333–81). San Francisco: Jossey-Bass.

Chase, W.G. and Simon, H.A. (1973). Perception in chess. *Cognitive Psychology*, 4:1, pp. 55–81.

Collier, S.G. and Folleso, K. (1995). SACRI: A measure of situation awareness for nuclear power plant control rooms. Proceedings of an International Conference: Experimental Analysis and Measurement of Situation Awareness (pp. 115–22). Daytona Beach, FL.

Cowans, N. (1988). Evolving conceptions of memory storage, selective attention, and their mutual constraints within the human information processing system. *Psychological Bulletin*, 104:2, pp. 163–91.

Crandall, B., Klein, G. and Hoffman, R. (2006). *Working minds: A Practitioner's Guide to Cognitive Task Analysis*. Cambridge, MA: MIT Press.

Dekker, A.H. (2003). Using agent-based modelling to study organisational performance and cultural differences. Proceedings of the MODSIM 2003 International Congress on Modelling and Simulation, Townsville, Queensland, 1793–1798. Available at www.mssanz.org.au/modsim03/Media/Articles/Vol 4 Articles/1793-1798.pdf.

Dennehy, K. (1997). Cranfield – Situation Awareness Scale: User Manual. Applied Psychology unit, College of Aeronautics, Cranfield University, COA report No. 9702, Bedford, January 1997.

Durso, F.D. and Gronlund, S.D. (1999). Situation awareness. In F.T. Durso, R. Nickerson, R. Schvaneveldt, S. Dumais, M. Chi and S. Lindsay (eds), *The Handbook of Applied Cognition* (pp. 283–314). London: Wiley.

Durso, F.T., Hackworth, C A., Truitt, T., Crutchfield, J. and Manning, C.A. (1998). Situation awareness as a predictor of performance in en route air traffic controllers. *Air Traffic Quarterly*, 6:1, pp. 1–20.

Dominguez, C. (1994). Can SA be defined? In M. Vidulich, C. Dominguez, E. Vogel and G. McMillan (eds), *Situation Awareness: Papers and Annotated Bibliography*. Report AL/CF-TR-1994-0085). Wright-Patterson Airforce Base, OH: Air Force Systems Command.

Edgar, G.K. and Edgar, H.E. (2007). Using signal detection theory to measure situation awareness: the technique, the tool (QUASA), the test, the way forward. In J. Noyes and M. Cook (eds), *Decision making in Complex Environments* (pp. 373–85). Aldershot: Ashgate Publishing.

Embrey, D.E. (1986). SHERPA: A systematic human error reduction and prediction approach. Paper presented at the International Meeting on Advances in Nuclear Power Systems, Knoxville, Tennessee.

Endsley, M.R. (1989). Final report: Situation awareness in an advanced strategic mission (NOR DOC 89–32). Hawthorne, CA: Northrop Corporation.

Endsley, M.R. (1990). Predictive utility of an objective measure of situation awareness. In Proceedings of the Human Factors Society 34th Annual Meeting, pp. 41–5. Santa Monica, CA: Human Factors Society.

Endsley, M.R. (1993). A survey of situation awareness requirements in air-to-air combat fighters. *The International Journal of Aviation Psychology*, 3:2, pp. 157–68.

Endsley, M.R. (1995a). Towards a theory of situation awareness in dynamic systems. *Human Factors*, 37:1, pp. 32–64.

Endsley, M.R. (1995b). Measurement of situation awareness in dynamic systems. *Human Factors*, 37:1, pp. 65–84.

Endsley, M.R. (2000). Theoretical underpinnings of situation awareness: A critical review. In M.R. Endsley and D.J. Garland (eds), *Situation Awareness Analysis and Measurement*, Mahwah, NJ: Laurence Erlbaum Associates.

Endsley, M.R. (2001). Designing for situation awareness in complex systems. In Proceedings of the Second International Workshop on Symbiosis of Humans, Artifacts and Environment, Kyoto, Japan.

Endsley, M.R. (2004). Situation awareness: Progress and directions. In S. Banbury and S. Tremblay (eds), *A Cognitive Approach to Situation Awareness: Theory, Measurement and Application* (pp. 317–41). Aldershot: Ashgate Publishing.

Endsley, M. R. and Jones, W. M. (1997). Situation awareness, information dominance, and information warfare. Technical Report 97-01. Belmont, MA: Endsley Consulting.

Endsley, M.R. and Jones, W. M. (2001). A model of inter and intra team situation awareness: Implications for design, training and measurement. In M. McNeese, E. Salas and M. Endsley (eds), *New Trends in Cooperative Activities: Understanding System Dynamics in Complex Environments*. Santa Monica, CA: Human Factors and Ergonomics Society.

Endsley, M.R. and Kaber, D.B. (1999). Level of automation effects on performance, situation awareness and workload in a dynamic control task. *Ergonomics*, 42:3, pp 462–92.

Endsley, M.R. and Kiris, E.O. (1995). *Situation Awareness Global Assessment Technique (SAGAT) TRACON Air Traffic Control Version User Guide*. Lubbock TX: Texas Tech University.

Endsley, M.R. and Robertson, M.M. (2000). Situation awareness in aircraft maintenance teams. *International Journal of Industrial Ergonomics*, 26:2, pp. 301–25.

Endsley, M.R. and Smolensky, M. (1998). Situation awareness in air traffic control: The picture. In M. Smolensky and E. Stein (eds), *Human Factors in Air Traffic Control* (pp. 115–54). New York: Academic Press.

Endsley, M.R., Bolte, B. and Jones, D. G. (2003). *Designing for Situation Awareness: An Approach to User-centred Design*. London: Taylor and Francis.

Endsley, M.R., Sollenberger, R. and Stein, E. (2000). Situation awareness: A comparison of measures. In Proceedings of the Human Performance, Situation Awareness and Automation: User-Centered Design for the New Millennium. Savannah, GA: SA Technologies, Inc.

Endsley, M.R., Selcon, S.J., Hardiman, T.D. and Croft, D.G. (1998). A comparative evaluation of SAGAT and SART for evaluations of situation awareness. In

Proceedings of the Human Factors and Ergonomics Society Annual Meeting (pp. 82–6). Santa Monica, CA: Human Factors and Ergonomics Society.

Endsley, M.R., Holder, C.D., Leibricht, B.C., Garland, D.C., Wampler, R.L. and Matthews, M.D. (2000). *Modeling and Measuring Situation Awareness in the Infantry Operational Environment. (1753)*. Alexandria, VA: Army Research Institute.

Entin, E. and Entin, E. (2000). Assessing team situation awareness in simulated military missions. Ergonomics for the New Millennium. In Proceedings of the XIVth Triennial Congress of the International Ergonomics Association and 44th Annual Meeting of the Human Factors and Ergonomics Society, San Diego, CA, Vol. 1, pp. 73–6.

Eysenck, M.W. and Keane, M.T. (1990). *Cognitive Psychology: A Student's Handbook*. Hove: Lawrence Erlbaum.

Farley, T.C., Hansman, R.J., Amonlirdviman, K. and Endsley, M.R. (2000). Shared information between pilots and controllers in tactical air traffic control. *Journal of Guidance, Control and Dynamics*, 23:5, pp. 826–36.

Fiore, S.M., Salas, E., Cuevas, H.M. and Bowers, C.A. (2003). Distributed coordination space: Toward a theory of distributed team process and performance. *Theoretical Issues in Ergonomics Science*, 4:3–4, pp. 340–64.

Fleishman, E.A. and Zaccaro, S.J. (1992). Toward a taxonomy of team performance functions. In R.W. Swezey and E. Salas (eds), *Teams: Their Training and Performance* (pp. 31–56). Norwood, NJ: Ablex.

Fox, J., Code, S.L. and Langfield-Smith, K. (2000). Team mental models: Techniques, methods and analytic approaches. *Human Factors*, 42:2, pp. 242–71.

Fracker, M. (1991). Measures of situation awareness: Review and future directions (Report No. AL-TR-1991-0128). Wright Patterson Air Force Base, OH: Armstrong Laboratories, Crew Systems Directorate.

Gorman, J.C., Cooke, N. and Winner, J.L. (2006). Measuring team situation awareness in decentralised command and control environments. *Ergonomics*, 49:12–13, pp. 1312–26.

Green, D., Stanton, N.A., Walker, G.H. and Salmon, P.M. (2005). Using wireless technology to develop a virtual reality command and control centre. *Virtual Reality*, 8:3, pp. 147–55.

Gregoriades, A. and Sutcliffe A. (2008). Workload prediction for improved design and reliability of complex systems. *Reliability Engineering and System Safety*, 93:4, pp. 530–49.

Gugerty, L.J. (1997). Situation awareness during driving: Explicit and implicit knowledge in dynamic spatial memory. *Journal of Experimental Psychology: Applied*, 3:1, pp. 42–66.

Hazlehurst, B., McMullen, C.K. and Gorman, P.N. (2007). Distributed cognition in the heart room: How situation awareness arises from coordinated communications during cardiac surgery. *Journal of Biomedical Informatics*, 40:5, pp. 539–51.

Hart, S.G. and Staveland, L.E. (1988). Development of a multi-dimensional workload rating scale: Results of empirical and theoretical research. In P.A. Hancock and N. Meshkati (eds), *Human Mental Workload*. Amsterdam: Elsevier.

Hauss, Y. and Eyferth, K. (2003). Securing future ATM-concepts' safety by measuring situation awareness in ATC. *Aerospace Science and Technology*, 7:6, pp. 417–27.

Head, H. (1920). *Studies in Neurology*, Vol. I. London: Oxford University Press.

Helmreich, R.L. and Foushee, H. C. (1993). Why crew resource management? Empirical and theoretical bases of human factors training in aviation. In E. Wiener, B. Kanki, and R. Helmreich (eds), *Cockpit Resource Management* (pp. 3–45). San Diego, CA: Academic Press.

Hogg, D.N., Folleso, K., Strand-Volden, F. and Torralba, B. (1995). Development of a situation awareness measure to evaluate advanced alarm systems in nuclear power plant control rooms. *Ergonomics*, 38:11, pp. 2394–413.

Hole, G. (2007). *The Psychology of Driving*. Mahwah, NJ: Lawrence Erlbaum Associates.

Hollnagel, E. (1993). *Human Reliability Analysis: Context and Control*. London: Academic Press.

Hollnagel, E. (1998). *Cognitive Reliability and Error Analysis Method (CREAM)*. New York: Elsevier.

Hollnagel, E. (2001). Extended cognition and the future of ergonomics. *Theoretical Issues in Ergonomics Science*, 2:3, pp. 309–15.

Hollnagel, E. and Woods, D.D. (1999). Cognitive systems engineering: New wine in new bottles. *International Journal of Human-Computer Studies*, 51:1–2, pp. 339–56.

Houghton, R.J., Baber, C., McMaster, R., Stanton, N.A., Salmon, P. M., Stewart, R. and Walker, G.H. (2006). Command and control in emergency services operations: A social network analysis. *Ergonomics*, 49, pp. 1204–25.

Hourizi, R. and Johnson, P. (2003). Towards an explanatory, predictive account of awareness. *Computers and Graphics*, 27, pp. 859–72.

Hutchins, E. (1995). *Cognition in the Wild*. Cambridge, MA: MIT Press.

Hutchins, E. and Klausen, T. (1996). Distributed cognition in an airline cockpit. In D. Middleton and Y. Engeström (eds), *Communication and Cognition at Work*. Cambridge: Cambridge University Press.

James, N. and Patrick, J. (2004). The role of situation awareness in sport. In S. Banbury and S. Tremblay (eds), *A Cognitive Approach to Situation Awareness: Theory and Application* (pp. 297–316). Aldershot: Ashgate Publishing.

Jeannott, E., Kelly, C. and Thompson, D. (2003). The development of situation awareness measures in ATM systems. EATMP report. HRS/HSP-005-REP-01.

Jenkins, D.P., Stanton, N.A., Walker, G.H., Salmon, P.M and Young, M.S. (2008) Applying cognitive work analysis to the design of rapidly reconfigurable interfaces in complex networks. *Theoretical Issues in Ergonomics Science*, 9, pp. 273–95.

Jensen, E. (2007). Sensemaking in military planning: A methodological study of command teams. *Cognition, Technology and Work*, 11:2, pp. 103–18.

Jones, D.G. and Endsley, M.R. (2000). Can real-time probes provide a valid measure of situation awareness? Proceedings of the Human Performance, Situation Awareness and Automation: User Centred Design for the New Millennium Conference, October 2000.

Jones, D.G. and Kaber, D.B. (2004). Situation awareness measurement and the situation awareness global assessment technique. In N. Stanton, Hedge, Hendrick, K. Brookhuis and E. Salas (eds), *Handbook of Human Factors and Ergonomics Methods* (pp. 42.1–42.7) Boca Raton, FL: CRC Press.

Kaber, D.B. and Endsley, M.R. (1998). Team situation awareness for process control safety and performance. *Process Safety Progress*, 17, pp. 43–8.

Kaber, D.B., Perry, C.M., Segall, N., McClernon, C.K. and Prinzel, L.P. (2006). Situation awareness implications of adaptive automation for information processing in an air traffic control-related task. *International Journal of Industrial Ergonomics*, 36, pp. 447–62.

Kerr, J. S. (1991). Driving without attention mode (DWAM): A normalisation of inattentive states in driving. In *Vision in Vehicles III.* North Holland: Elsevier.

Kirwan, B. (1992). Human error identification in human reliability assessment. Part 2: Detailed comparison of techniques. *Applied Ergonomics*, 23, pp. 371–81.

Kirwan, B. (1998). Human error identification techniques for risk assessment of high-risk systems – Part 1: Review and evaluation of techniques. *Applied Ergonomics*, 29, pp. 157–77.

Klein, G. (1998). Sources of power: How people make decisions, MIT Press, Cambridge, MA.

Klein, G. (2000). Cognitive task analysis of teams. In J.M. Schraagen, S.F. Chipman, and V.L. Shalin (eds), *Cognitive Task Analysis*. Mahwah, NJ: Lawrence Erlbaum, pp. 417–30.

Klein, G. and Armstrong, A.A. (2004). Critical decision method. In N.A. Stanton, A. Hedge, E. Salas, H. Hendrick and K. Brookhaus (eds), *Handbook of Human Factors and Ergonomics Methods* (pp. 35.1–35.8), Boca Raton, FL: CRC Press,

Klein, G. and Miller, T.E. (1999). Distributed planning teams. *International Journal of Cognitive Ergonomics*, 3, pp. 203–22.

Kuper, S.R. and Giurelli, B.L. (2007). Custom work aids for distributed command and control: A key to enabling highly effective teams. *The International Command and Control Journal*, 1, pp. 24–42.

Lawson, J.S. (1981) Command and control as a process. *IEEE Control Systems Magazine*, March, pp. 86–93.

Lloyd, M. and Alston, A. (2003). Shared awareness and agile mission groups, 8th ICCRTS, NDU Washington DC. http://www.dodccrp.org/8thICCRTS/Pres/plenary/3_0830Lloyd.pdf.

Ma, R. and Kaber, D.B. (2007). Situation awareness and driving performance in a simulated navigation task. *Ergonomics*, 50, pp. 1351–64.

Macey, P. (2007). Critical Success Factors. Command and Control Development Centre (C2DC) unpublished internal report.

Matthews, M.D. and Beal, S.A. (2002). Assessing situation awareness in field training exercises. US Army Research Institute for the Behavioural and Social Sciences. Research Report 1795.

Matthews, M.D., Strater, L.D. and Endsley, M.R. (2004). Situation awareness requirements for infantry platoon leaders. *Military Psychology*, 16, pp. 149–61.

Matthews, M.D., Pleban, R.J., Endsley, M.R. and Strater, L.D. (2000). Measures of infantry situation awareness for a virtual MOUT environment. Proceedings of the Human Performance, Situation Awareness and Automation: User Centred Design for the New Millennium Conference, October 2000.

May, J.L. and Gale, A.G. (1998). How did I get here? Driving without attention mode. In M. Hanson (ed.), *Contemporary Ergonomics*. London: Taylor and Francis.

McGuinness, B. and Ebbage, L. (2002). Assessing human factors in command and control: Workload and situational awareness metrics. In Proceedings of the 2002 Command and Control Research and Technology Symposium, Monterey, CA, 2002.

McGuinness, B. and Foy, L. (2000). A subjective measure of SA: The crew awareness rating scale (CARS). Presented at the Human Performance, Situational Awareness and Automation Conference, Savannah, GA, 16–19 October 2000.

Ministry of Defence (2004). Board of inquiry report into Challenger II incident. http://www.mod.uk/DefenceInternet/AboutDefence/CorporatePublications/ BoardsOfInquiry/BoardOfInquiryIntoTheChallenger2Incident25Mar03.htm.

Ministry of Defence (2005). Joint service publication 777 – Network enabled capability. Version 1, Edition 1. http://www.mod.uk/DefenceInternet/AboutDefence/Corporate Publications/ScienceandTechnologyPublications/NEC/.

Ministry of Defence (2007). *The Combat Estimate, Combined Armed Staff Trainer (CAST) Guide*. Warminster: Land Warfare Centre.

Morgan, B.B., Glickman, A.S., Woodard, E.A., Blaiwes, A.S. and Salas, E. (1986). Measurement of team behavior in a Navy training environment (Tech. Report TR-86-014). Orlando, FL: Naval Training Systems Center, Human Factors Division.

Moray, N. (2004). Ou' sont les neiges d'antan? (Where are the snows of yesterday?) Proceedings of the HPSAAII. Mahwah, NJ: LEA.

NATO (2002). Code of best practice for C2 assessment. Department of Defense Command and Control Research Program (CCRP), 3rd edition.

Neisser, U. (1976). *Cognition and Reality: Principles and Implications of Cognitive Psychology*. San Francisco: Freeman.

Nofi, A. (2000). Defining and Measuring Shared Situational Awareness, DARPA. http:// www.thoughtlink.com/publications/DefiningSSA00Abstract.htm.

Norman, D.A. (1981). Categorization of action slips. *Psychological Review*, 88, pp. 1–15.

Norman, D.A. and Shallice, T. (1986). Attention to action: Willed and automatic control of behavior. In R.J. Davidson, G.E. Schwartz and D. Shapiro (eds), *Consciousness and Self-regulation: Advances in Research*, Vol. IV. New York: Plenum Press.

O'Hare, D., Wiggins, M., Williams, A. and Wong, W. (2000). Cognitive task analysis for decision centred design and training. In J. Annett and N.A. Stanton (eds), *Task Analysis* (pp. 170–90). London: Taylor and Francis.

Orasanu, J. and Fischer, U. (1997). Finding decisions in natural environments: The view from the cockpit. In C.E. Zsambok and G. Klein (eds), *Naturalistic Decision Making* (pp. 343–57). Mahwah, NJ: Lawrence Erlbaum.

Ottino, J. (2003). Complex systems. *AIChE Journal*, 49, pp. 292–9.

Paris C., Salas E. and Canon-Bowers J. A. (2000). Teamwork in multi-person systems: A review and analysis. *Ergonomics*, 43, pp. 1052–75.

Patrick, J., James, N., Ahmed, A. and Halliday, P. (2006). Observational assessment of situation awareness, team differences and training implications. *Ergonomics*, 49, pp 393–417.

Perla, P., Markowitz, M., Nofi, A., Weuve, C., Loughran, J. and Stahl, M. (2000). Gaming and shared situation awareness. DARPA, http://www.thoughtlink.com/publications/GamingSSA00Abstract.html.

Piaget, J. (1926). *The Child's Conception of the World*. New York: Harcourt, Brace.

Rasker, P.C., Post, W.M. and Schraagen, M.C. (2000). Effects of two types of intra-team feedback on developing a shared mental model in command and control teams. *Ergonomics*, 43, pp. 1167–89.

Reber, A. (1995). *Implicit Learning and Tacit Knowledge: An Essay on the Cognitive Unconscious*. New York: Oxford University Press.

Riley, J.M., Endsley, M.R., Bolstad, C.A. and Cuevas, H.M. (2006). Collaborative planning and situation awareness in army command and control. *Ergonomics*, 49, pp. 1139–53.

Rognin, L., Salembier, P. and Zouinar, M. (1998). Cooperation, interactions and socio-technical reliability: The case of air-traffic control. Comparing French and Irish settings. Proceedings of ECCE 9, pp. 19–24.

Rousseau, R., Tremblay, S. and Breton, R. (2004). Defining and modeling situation awareness: A critical review. In S. Banbury and S. Tremblay (eds), *A Cognitive Approach to Situation Awareness: Theory and Application* (pp. 3–21). Aldershot: Ashgate Publishing.

Salas, E. (2004). Team methods. In N.A. Stanton, A. Hedge, K. Brookhuis, E. Salas, and H. Hendrick (eds), *Handbook of Human Factors and Ergonomics Methods* (pp. 43-1–43-4). Boca Raton, FL: CRC Press.

Salas, E., Burke, C.S. and Samman, S.N. (2001). Understanding command and control teams operating in complex environments. *Information Knowledge Systems Management*, 2, pp. 311–23.

Salas, E., Muniz, E.J. and Prince, C. (2006). Situation awareness in teams. In W. Karwowski (ed.), *International Encyclopedia of Ergonomics and Human factors*, Vol. 1 (pp. 903–6). Boca Raton, FL: Taylor and Francis.

Salas, E., Sims, D.E. and Burke, C.S. (2005) Is there a big five in teamwork? *Small Group Research*, 36, pp. 555–99.

Salas, E., Stout, R.J. and Cannon-Bowers, J.A. (1994). The role of shared mental models in developing shared situational awareness. In Gilson, D., Garland, D.J. and Koonce, J.M. (eds), *Situational Awareness in Complex Systems: Proceedings of a CAHFA Conference*. Daytona Beach, FL: Embry-Riddle Aeronautical University Press.

Salmon, P.M., Jenkins, D.P., Stanton, N.A., Walker and G.H. (2007). ComBAT evaluation: WESTT analysis and usability study results. Human Factors Integration Defence Technology Centre report.

Salas, E., Prince, C., Baker, P.D. and Shrestha, L. (1995). Situation awareness in team performance. *Human Factors*, 37, pp. 123–36.

Salmon, P., Stanton, N., Walker, G. and Green, D. (2006). Situation awareness measurement: A review of applicability for C4i environments. *Journal of Applied Ergonomics*, 37, pp. 225–38.

Salmon, P.M, Stanton, N.A., Walker, G.H., Baber, C., Jenkins, D.P. and McMaster, R. (2008a). What really is going on? Review of situation awareness models for individuals and teams. *Theoretical Issues in Ergonomics Science*, 9, pp. 297–323.

Salmon, P.M., Stanton, N.A., Walker, G.H., Jenkins, D.P., Baber, C. and McMaster, R. (2008b). Representing situation awareness in collaborative systems: A case study in the energy distribution domain. *Ergonomics*, 51, pp. 367–84.

Sarter, N.B. and Woods, D.D. (1991). Situation awareness – a critical but ill-defined phenomenon. *International Journal of Aviation Psychology*, 1, pp. 45–57.

Shu, Y. and Furuta, K. (2005). An inference method of team situation awareness based on mutual awareness. *Cognition Technology and Work*, 7, pp. 272–87.

Shadbolt, N.R. and Burton, M. (1995). Knowledge elicitation: A systemic approach. In J.R. Wilson and E.N. Corlett (eds), *Evaluation of Human Work: A Practical Ergonomics Methodology*, pp. 406–40.

Siemieniuch, C.E. and Sinclair, M.A. (2006). Systems integration. *Journal of Applied Ergonomics*, 37, pp. 91–110.

Smith, E.A. (2002). Effects based operations: Applying network centric warfare in peace, crisis, and war. Department of Defense Command and Control Research Program (CCRP) Report, http://www.dodccrp.org/html3/pubs_download.html.

Smith, K. and Hancock, P.A. (1995). Situation awareness is adaptive, externally directed consciousness. *Human Factors*, 37, pp. 137–48.

Smolensky, M.W. (1993). Toward the physiological measurement of situation awareness: The case for eye movement measurements. In Proceedings of the Human Factors and Ergonomics Society 37th Annual Meeting, Santa Monica, Human Factors and Ergonomics Society.

Sonnenwald, D.H., Maglaughlin, K.L. and Whitton, M.C. (2004). Designing to support situation awareness across distances: An example from a scientific collaboratory. *Information Processing and Management*, 40, pp. 989–1011.

Stanton, N.A. (2006). Hierarchical task analysis: Developments, applications and extensions. *Applied Ergonomics*, 37, pp. 55–79.

Stanton, N.A. and Stevenage, S.V. (1998). Learning to predict human error: Issues of reliability, validity and acceptability. *Ergonomics*, 41, pp. 1737–56.

Stanton, N.A. and Young, M. S. (1999a). What price ergonomics? *Nature*, 399, pp. 197–8.

Stanton, N.A and Young, M. S. (1999b). *A Guide to Methodology in Ergonomics: Designing for Human Use*, London: Taylor and Francis.

Stanton, N.A. and Young, M.S. (2000). A proposed psychological model of driving automation. *Theoretical Issues in Ergonomics Science*, 1:4, pp. 315–31.

Stanton, N.A. and Young, M.S. (2003). The application of ergonomics methods by novices. *Applied Ergonomics*, 34, pp. 479–90.

Stanton, N.A., Chambers, P.R.G. and Piggott, J. (2001). Situational awareness and safety. *Safety Science*, 39, pp. 189–204.

Stanton, N., Salmon, P.M., Walker, G.H. and Jenkins, D.P. (2009). Genotype and phenotype schema and their role in distributed situation awareness in collaborative systems. *Theoretical Issues in Ergonomics Science*, 10:1, pp. 43–68.

Stanton, N.A., Hedge, A., Brookhuis, K., Salas, E. and Hendrick, H. (eds) (2004). *Handbook of Human Factors Methods*. Boca Raton, FL: CRC Press.

Stanton, N.A., Salmon, P.M., Walker, G., Baber, C. and Jenkins, D.P. (2005). *Human factors Methods: A Practical Guide for Engineering and Design*. Aldershot: Ashgate Publishing.

Stanton, N.A., Walker, G.H., Jenkins, D.P., Salmon, P.M., Rafferty, L. and Revell, K. (2008). OFT 3: Conclusions and recommendations. Unpublished HFI-DTC Report, April 2008.

Stanton, N.A., Stewart, R., Harris, D., Houghton, R.J., Baber, C., McMaster, R., Salmon, P.M., Hoyle, G., Walker, G.H., Young, M.S., Linsell, M., Dymott, R. and Green, D. (2006). Distributed situation awareness in dynamic systems: Theoretical development and application of an ergonomics methodology. *Ergonomics*, 49, pp 1288–311.

Stewart, R., Stanton, N.A., Harris, D., Baber, C., Salmon, P.M., Mock, M., Tatlock, K., Wells, L. and Kay, A. (2008). Distributed situation awareness in an airborne warning and control system: Application of novel ergonomics methodology. *Cognition Technology and Work*, 10:3, pp. 221–29.

Stone, N.J. and Posey, M. (2008). Understanding coordination in computer-mediated versus face-to-face groups. *Computers in Human Behavior*, 24, pp. 827–51.

Stout, R.J., Cannon-Bowers, J.A., Salas, E. and Milanovich, D.M. (1999). Planning, shared mental models, and coordinated performance: An empirical link is established. *Human Factors*, 41, pp. 61–71.

Taylor, R.M. (1990). Situational awareness rating technique (SART): The development of a tool for aircrew systems design. In *Situational Awareness in Aerospace Operations (AGARD-CP-478)*, pp. 3/1–3/17. Neuilly Sur Seine, France: NATO-AGARD.

Uhlarik, J. and Comerford, D.A. (2002). A review of situation awareness literature relevant to pilot surveillance functions. (DOT/FAA/AM-02/3) Washington, DC: Federal Aviation Administration, U.S. Department of Transportation.

US Army (1994). Multiservice procedures for humanitarian assistance operations. Multiservice tactics, techniques, and procedures. Field Manual 100-23-1, Air Land Sea application center, October 1994.

USJFCOM (2005). Effects-based approach to multinational operations: Concept of operations (CONOPS) with implementing procedures. Version 0.85, 10 November 2005, United States Joint Forces Command Joint Experimentation Directorate.

Vidulich, M.A. and Hughes, E.R. (1991). Testing a subjective metric of situation awareness. Proceedings of the Human Factors Society 35th Annual meeting (pp. 1307–11). Santa Monica, CA: Human Factors Society.

Waag, W.L. and Houck, M.R. (1994). Tools for assessing situational awareness in an operational fighter environment. *Aviation, Space and Environmental Medicine*, 65, pp. A13–A19.

Walker, G.H. (2004). Verbal protocol analysis. In N.A. Stanton, A. Hedge, K. Brookhuis, E. Salas, and H. Hendrick. (eds), *Handbook of Human Factors Methods*. Boca Raton, FL: CRC Press.

Walker. G.H., Gibson, H., Stanton, N.A., Baber, C., Salmon, P.M. and Green, D. (2006a). Event analysis of systemic teamwork (EAST): A novel integration of ergonomics methods to analyse C4i activity. *Ergonomics*, 49, pp 1345–69.

Walker, G.H., Stanton, N.A., Kazi, T.A., Salmon, P. and Jenkins (2009). Does advanced driver training improve situation awareness? *Applied Ergonomics*, 40:4, 678–87. IF 0.94.

Walker. G.H., Stanton, N.A., Stewart, R., Jenkins, D.P. Wells, L., McMaster, R. and Ellis, A. (2006b). CAST evaluation. Unpublished report.

Weick, K.E. (1993). The collapse of sensemaking in organizations: The Mann Gulch disaster. *Administrative Science Quarterly*, 38, pp. 628–52.

Wellens, A.R. (1993). Group situation awareness and distributed decision-making: From military to civilian applications. In N.J. Castellan (ed.), *Individual and Group Decision Making: Current Issues* (pp. 267–87). Hillsdale, NJ: Erlbaum Associates.

Whalley, S.P. and Kirwan, B. (1989). An evaluation of five human error identification techniques. Paper presented at the fifth International Loss Prevention Symposium, Oslo, June 1989.

Wilson, K. A., Salas, E., Priest, H. A. and Andrews, D. (2007). Errors in the heat of battle: Ttaking a closer look at shared cognition breakdowns through teamwork. *Human Factors*, 49, pp. 243–56.

Wood, D. (1998). How Children Think and Learn (2nd edition). Oxford: Blackwell Publishing.

Young, J.E, Klosko, J.S. and Weishaat, M.E. (2003). *Schema Therapy: A Practitioner's Guide*. New York: The Guildford Press.

Index